Studies in Systems, Decision and Control

Volume 181

Series editor

Janusz Kacprzyk, Polish Academy of Sciences, Warsaw, Poland
e-mail: kacprzyk@ibspan.waw.pl

The series "Studies in Systems, Decision and Control" (SSDC) covers both new developments and advances, as well as the state of the art, in the various areas of broadly perceived systems, decision making and control–quickly, up to date and with a high quality. The intent is to cover the theory, applications, and perspectives on the state of the art and future developments relevant to systems, decision making, control, complex processes and related areas, as embedded in the fields of engineering, computer science, physics, economics, social and life sciences, as well as the paradigms and methodologies behind them. The series contains monographs, textbooks, lecture notes and edited volumes in systems, decision making and control spanning the areas of Cyber-Physical Systems, Autonomous Systems, Sensor Networks, Control Systems, Energy Systems, Automotive Systems, Biological Systems, Vehicular Networking and Connected Vehicles, Aerospace Systems, Automation, Manufacturing, Smart Grids, Nonlinear Systems, Power Systems, Robotics, Social Systems, Economic Systems and other. Of particular value to both the contributors and the readership are the short publication timeframe and the world-wide distribution and exposure which enable both a wide and rapid dissemination of research output.

More information about this series at http://www.springer.com/series/13304

Alla G. Kravets

Editor

Big Data-driven World: Legislation Issues and Control Technologies

Springer

Editor
Alla G. Kravets
CAD&Research Department
Volgograd State Technical University
Volgograd, Russia

ISSN 2198-4182   ISSN 2198-4190   (electronic)
Studies in Systems, Decision and Control
ISBN 978-3-030-13174-6   ISBN 978-3-030-01358-5   (eBook)
https://doi.org/10.1007/978-3-030-01358-5

This Springer imprint is published by the registered company Springer Nature Switzerland AG
The registered company address is: Gewerbestrasse 11, 6330 Cham, Switzerland

Preface

Big Data-driven world is a new reality which defines modern opportunities, relations, and, of course, challenges. In this book, researchers from the scientific and educational organizations attempt to study the phenomena of Big Data-driven world with the comprehensive approach. This approach combines traditional IT methodologies and the overview of juridical and legislative rules in domains of Big Data, Internet, and IT.

The first part of the book "Theoretical Concepts of Control Technologies in Big Data-driven World" defines methodological foundations of Big Data-driven world, formulates its concept within the frameworks of modern control methods and theories, and views the peculiarities of Control Technologies as a specific sphere of Big Data-driven world, distinguished in the modern Digital Economy. The authors study the genesis of mathematical and information methods transition from data analysis and processing to knowledge discovery and predictive analytics in the twenty-first century.

The second part "The Methodological Framework of Legislation Regulation in Big Data-driven World" is devoted to studying Big Data-driven world through the prism of legislation issues. The authors determine the legislative foundations of the Big Data-driven world concept as a breakthrough in the modern information technologies. The chapter also analyzes the conditions of development and implementation of Big Data analysis approaches in the investigative activities and determines the role and meaning of global networks as platforms for the establishment of legislation regulation in Big Data-driven world.

In the third part "Counteraction of Terrorism and Extremism Challenges in Big Data-driven World", the authors substantiate the scientific and methodological approaches to study modern mechanisms of terrorism and extremism counteraction in the Big Data-driven world. Internet technologies defined new challenges of dissemination and accessibility of socially dangerous information. The authors determine the main features of extremist information and financing flows and the key activities of counteraction and offer criteria for evaluating the effectiveness of software and hardware solutions.

Fourth part "Practical Aspects and Case-Studies of Legislation Regulation and Control Technologies Development" is devoted to the analysis of the accumulated experience of formation and development of Big Data solutions in the legislative and control Russian and international practice. The authors perform systematization of successful experience of the Big Data solutions establishment in the different domains and analyze causal connections of the Digital Economy formation from the positions of new technological challenges.

Volgograd, Russia   Alla G. Kravets
July 2018

Contents

Part I
Theoretical Concepts of Control Technologies in Big Data-driven World

Methodological Foundations of the Digital Economy

Dmitry Novikov and Mikhail Belov

Abstract The complex activity of human is a fundamental element of any economy, including a digital one. In connection with the above, the development of methodological aspects of integrated activities is an urgent task. The methodological basis of the research is a system analysis, as well as the theory of multi-agent systems. The main trends distinguishing the "digital" economy from "non-digital" one are highlighted based on the systematic approach. The concept of complex activity (CA) is defined and the ontology of the basic concepts of CA methodology is offered. An integrated and coordinated system of CA models is introduced: the basic model of the structural element of activity (SEA), which is the main atomic element of CA representation and analysis; logical model of CA, reflecting the structure of the objectives of the activity and the hierarchy of subordination of elements of the CA; causal model of CA, describing the cause-effect relationship between the elements of CA and, in fact, CA technology; process model that reflects the life cycle of activity and its elements. The role of information in the process of implementing the CA is considered. Significant factors of the role of information (information model) in the CA technology and, as a consequence, in the economy, are revealed. The chapter shows the possibility of constructive representation of complex activities in the form of multi-agent models. The stated results create the methodological bases for the analysis and construction of the digital economy as an integrated system.

Keywords Digital economy · Complex activity · Digital interaction

D. Novikov (✉)
V. A. Trapeznikov Institute of Control Sciences, Russian Academy of Sciences,
Moscow 117342, Russian Federation
e-mail: novikov@ipu.ru

M. Belov
IBS Company, Moscow 127018, Russian Federation
e-mail: mbelov@ibs.ru

© Springer Nature Switzerland AG 2019

A. G. Kravets (ed.), *Big Data-driven World: Legislation Issues and Control Technologies*, Studies in Systems, Decision and Control 181,
https://doi.org/10.1007/978-3-030-01358-5_1

1 Introduction

Considering the phenomenon of "digital economy", we'll rely on the definition given in [1], which refers to the need for accelerated development of the "digital economy of the Russian Federation, in which data in digital form is a key factor in production in all spheres of socio-economic activity, which raises the country's competitiveness, the quality of life of citizens, ensures economic growth and national sovereignty".

The cited definition is focused on two key factors: first, it deals with all spheres of socio-economic activity (state); secondly, it determines the role of data in digital form. Thus, it makes sense to define the "digital economy" as an activity (all spheres of socio-economic activity), in which the data in digital form is the key factor in its implementation. Hence, the distinctive feature of the "digital economy" from "non-digital" one, and of the condition of the corresponding transformation is the use of data in digital form as a key factor in realizing social and economic activity.

This definition makes it necessary to identify and analyze the trends already existing for the transformation of the "non-digital" economy into the "digital" one.

First, it is the emergence of new activities, the results of which exist only in digital form. This is the activity related to the creation and use of social networks, computer entertainment and other forms of digital interaction of individuals, yet unknown.

Secondly, it is the transformation of known activities related to the transformation of key factors into digital form. First and foremost, it is the wide development of computer design and modeling of industrial products—the transition from drawings and other technical documents in a "paper" format to information models of products, systems, facilities, services, etc.

Thirdly, it is the automation of existing activities, including using artificial intelligence technologies, which leads to the disappearance of entire professions as a result of replacing people with computers and/or robots. As the most striking example, it makes sense to note the almost complete disappearance of typographic workers, typists, telephone station operators. A similar exclusion of the professions of translators of foreign languages, drivers of cars, captains of ships and aircraft will occur with a high probability in coming years.

Thus, the study of market trends and analysis of the phenomenon of "digital economy" require a deeper consideration of all aspects of the socio-economic activity.

Socio-economic activities are characterized by a wide variety of professions: an entrepreneur developing a high-tech business (for example, Ilon Mask, 1a), Chief Designer (S.P. Korolev, 1b), a state/public figure (Peter I, 1c), operator of a gas station (2a), sorter of oranges (2b), or a worker in the final assembly shop of an aircraft manufacturing enterprise (2c).

Everyone will say that the activities of some are complex and diverse (1a–1в), and the activities of others are monotonous and routine (2a–2в). But the activities of the latter are also sometimes "complicated": no one will dare say that the cockpit assembly of modern aircraft (2c) is not a "complex" activity. What they have in common is that they are all examples of human activity. And what are the differences? How should we formally determine the similarities and differences of different activities?

A distinctive characteristic is the uncertainty of activity—the uncertainty of external environment, the uncertainty of the goals, technologies, and the behavior of the actor. In 2a–2c the activity is completely determined or conceived as completely determined: they react to the uncertainty of the external environment in the form of escalating the problem to a higher-level entity and are responsible for performing actions within the prescribed "top" technology. Unlike 1a, 1b or 1c, which react not only to the uncertainty of environmental factors but to the uncertainty of certain goals and technology. Their reaction is realized in the form of the organization of new activities, first of all, the creation of a new activity technology. They themselves are responsible for the final result of the activity, including the activities of the subordinate entities. And why does "non-determinism" of the activity of 1a–1c appear and in what? Next, everyone will say that the activities of the former (1a–1c) are not always "equally complex". How to separate these different in complexity activities of the same person? It is intuitively clear that not all elements of activity are "equally complex and uncertain". Moreover, the implementation of elements of activity with a high degree of complexity and uncertainty requires significantly higher costs (managerial and material resources, time costs) compared with routine (less complex and uncertain) elements.

And what about management and organization activities? Is it possible to define specifically and formally the essence of such activity? For example, what does the general director manage while considering and delegating the e-mail messages? Is it a complex activity? And when does the same general director sign the contract? To sign a contract is his responsibility, but he puts the signature almost formally on the assumption of previous discussions.

Which of these types and elements of activity can be digitized? What do you need to create for this?

Following the results of [2], we'll understand the concept of "complex activity" as the activity [3] possessing a non-trivial internal structure, with multiple and/or changing actor, technology, and role of the actor of activity in its target context. Due to the nature of complex activities (CA), it should be considered together with the actor which is implementing this activity (usually a complex organizational and technical system).

Accordingly, the theory used for analyzing the methodological foundations of the "digital economy" in the present chapter, is called the methodology of complex activity (MCA).

Complex activity is a complex system, having a non-trivial internal structure, multiple and/or changing actor, technology, and role of the actor in its target context. The following characteristics should be noted as its most significant ones:

- logical and causal structures of CA;
- the life cycle, as an essential factor in the implementation of CA;
- the relationship of elements of CAs and their actors;
- process and design types of CA elements;
- the uncertainty of CA;
- generation and existence of CA elements in time.

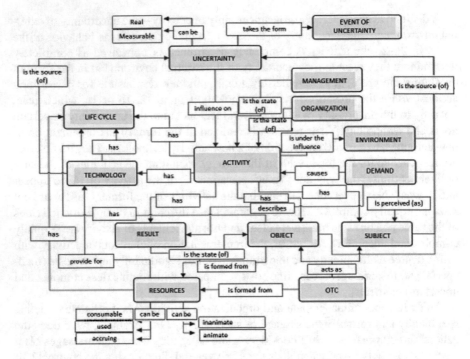

Fig. 1 Ontology of the basic MCA categories

The "ontology" of the basic categories of MCA is shown in Fig. 1 in the form of an "essence-relationship" diagram.

2 Models of Integrated Activities

The core of MCA is an integrated set of CA models, basic of which is the model of the structural element of activity (SEA). The SEA is defined as having the one (shown in Fig. 2) structure instance:

- formed to achieve a specific goal/obtaining a certain result (transformation of the actor of activity),
- characterizing the activity (aimed at obtaining a result) in accordance with a certain technology related to a certain actor,
- the actor of which are elements of some complex activity.

The SEA is designed to be used as a unified presentation formalism of a CA element model.

The arrows from the technology and from actions to the subject matter mean that the subject matter changes as a result of the actions of technology, from the subject

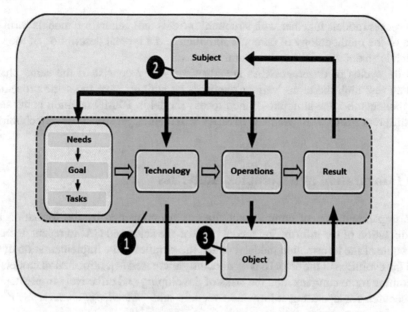

Fig. 2 Model of the structural element of activity

matter to the result, i.e. the result is the final state of the subject matter evolution in the process of the activity.

The structure of the element of complex activity (Fig. 2) is actually a composition of the scheme of the procedural components of activity and the scheme of the activity of its actor and subject matter.

Complex activity has different structural elements varying by their grounds, first of all, target and technological (cause-effect) ones. Therefore, corresponding models are needed to describe these structures.

Models of logical and causal structures of CA provide recursive integration of its elements, responding to fractal properties of complex activities.

The model of the logical structure of complex activities is built on the basis of the structure of the goals of the CA. The most productive is the use of the CA goals (and not its other components) as the basis for identifying the logical structure of the CA.

The logical structure of the CA reflects, among others, a hierarchy of management and responsibility links, while a cause-and-effect hierarchy reflects its technological links.

Activity (its elements), demand, subject matter, technology, resources are characterized by their own life cycle, therefore there is the topical problem of *coordinated management of life cycles* of demand, activity, its subject matter and actor, knowledge, technologies, and organizations.

The implementation of complex activities occurs, in the general case, in the form of generation, realization, and evolution of a set of interacting elements of a complex activity that form a hierarchical structure.

Process models together with structural models and generation models form the core of the methodology of complex activities and a typical description of the CA, which is an analog of the architectural template.

Life cycles of different SEAs are identical—they consist of the same phases and stages; and, therefore, various elements of activity have the same procedural components. So, the structure of the process model in BPMN notation is the same for different SEAs. The specificity of SEAs is in their target structure and technology.

3   Information and Complex Activities

The implementation of complex activities is accompanied by the formation and modification of the information model (IM) of the actor and CA. In recent decades, the value of the information model (created the parallel to the implementation of CA and the evolution of the actor) has significantly increased [4]. Information models are becoming more complex, and the tasks of developing and effectively implementing procedures for operating information models and *"knowledge management"* are becoming increasingly urgent.

There are many definitions for IM, for example, one of them is contained in the State Standard 34.003-90 [5]: the model of the subject matter, represented in the form of information describing the parameters and variables of the subject matter that are under consideration, the connections between them, the inputs and outputs of the subject matter, and allowing modeling possible states of the subject matter by submitting information on changes in input quantities.

It makes sense to note, that the IM has always existed and has always been used. For example, even in Ancient Egypt, the IMs of temples were first formed "in the heads" of architects, then they appeared in the form of primitive blueprints, and only then they were re-embodied in the form of structures preserved for many millennia. However, in recent decades the role of IM has significantly changed:

1. For a long time, the cost and timing of the implementation of the IM, that is, the creation of products were incommensurate with the cost and timing of the development of IM (now this ratio has changed). Accordingly, the share of the cost of raw materials in the cost of the finished product has decreased, and the share of the cost of IM development has increased. The cost of metal and composites, of which a modern car or aircraft is made, does not exceed tens of percent in the price of the product, the rest is the cost of the production of parts, final assembly and, actual design. The development of a new model aircraft lasts for 5–7 or more years, hundreds of engineers are involved in this activity, and the result is an immaterial IM. A vivid example in this case is the cost of the iPhone [6]: the cost of all components of the iPhone 5s (16 GB) is 191 dollars, another 8 dollars is spent on assembling devices, i.e. the total amount is 199 dollars, and the iPhone itself was sold in the US for $ 649 in 2014. So, a greater share in

the price, along with the value of the brand, takes design and promotion to the market, i.e., actually IM.

2. An information model, unlike a real product, exists at all stages of its life cycle, from concept to disposal.

3. Previously, IM was created in a single development center, and production could be implemented in cooperation with contractors and manufacturers. Now the process of product development, that is, the creation of IM is carried out in branched cooperation. For example, it is known that the development of Boeing CA aircraft is carried out by several engineering centers in the US, Australia, and Russia, similarly—the Airbus development.

4. In the past, information models in the form of drawings, specifications, and other technical documentation were used only for production, now it is not so. For example, practically from the very beginning of nuclear power development, the safety requirements for power units are confirmed on the basis of calculations, that is, on the basis of IM. In recent years, IM has also been used to certify cars, aircrafts and other technical facilities.

5. With the complication of production facilities and the expansion of the use of IM, the models themselves are becoming increasingly complex and expensive. Now the IMs contain not only a geometric description and structure of the product, materials and technological maps, logistical information, but also complex models of functioning, movements, and others. Today IM is a complex hierarchical multidisciplinary complex.

6. The process of separation of the "material" and "intellectual" parts of production is becoming ever more intense. An independent sub-sector of the "intellectual part" of production has been formed (creation, modernization of information models, etc.). In addition to the growth of the number of companies whose business is engineering services, new factors are emerging: first, the standardization of engineering services, and, as a consequence, "offshore developments"; and, secondly, a significant increase in the cost of the "intellectual part" of production/product in relation to the "material" one, primarily because of their complicated nature. So, the shift of emphasis from material subject matter to intellectual products is the reason for separating the "intellectual part" of production into an independent sub-sector.

7. "Intellectual part" of the product (i.e. the information model) is also institutionalized and increasingly becomes a commodity. The complete IM includes information components of marketing, requirements, results of design/construction, manufacturing technology, certification, logistics, plans, the structure of the cooperation of the manufacturer, and other attributes. IM is formed, changed and used at all stages of the life circle by all participants in the cooperation. The role of IM is enhanced, as it is used not only for the manufacture of a product (for example, as a basis for the design and technological organization of production) but also for the certification of products and other purposes.

In fact, complex activities have evolved into two parallel and interrelated processes:

(1) creation and maintenance (including modification) of the information model;
(2) implementation of actions related to the subject matter in accordance with this
     model, ensuring evolution during its life cycle, that is, actually the activity itself.

This transformation was an objective source of the revision of the role of information in the life of society, manifested in numerous discussions of "information
explosions", "transition to an information society", "digital economy", "knowledge
economy", etc.

The information model generally contains not only normative, a priori information
on complex activities, but also operational (in relation to specific subject matter) and
forecast, as well as various historical data, auxiliary information with a different level
of detail and formalization.

Complication of IM and the increase of its role objectively cause the need to
establish effective methods and tools for its creation, storage, use, modification,
maintenance of integrity and so on. These methods, procedures, and tools are the
subjects of several areas of knowledge and activities that are part of the broad "information technology" industry. The popular and widely discussed technologies of
Product Lifecycle Management (PLM) and *Knowledge Management* are of great
importance for the creation, maintenance and use of the integrated IT activities; see,
for example [7].

Product lifecycle management is defined [8] as a strategic business approach for
applying a consistent set of tools that support the joint creation, management, dissemination and use of product information within an extended enterprise, starting from
the product concept to the end of the life cycle, integrating employees, technological
processes, production systems and information. The methods and implementation of
PLM-class software are very widespread: virtually all modern production activity is
based on their use. According to the leading analytical agencies, the market for PLM
products is tens of billions of dollars per year and continues to grow rapidly [9].

Knowledge management is designed to operate on a broader and less formalized
spectrum of information. It is defined [10] as a discipline that develops an integrated
approach to the definition, collection, systematization, search and provision of all
enterprise information assets, which may include databases, documents, policies,
procedures, as well as the knowledge and experience of employees not yet collected.
The main tasks of knowledge management are the classification and structuring of
professional knowledge, the creation of databases and repositories of professional
experience and expertise, the organization of professional communities and information exchange within them, etc.

It can be said that if PLM tools and methods provide management of the IM of
the CA actor, then knowledge management is the management of the information
model of the CA as a whole.

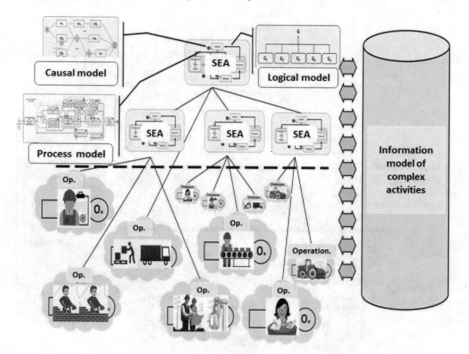

Fig. 3 System-wide structural and behavioral properties of the SEA

4 Model of the Process of Implementation of Complex Activities

Within the MCA, all SEAs have a single set of system-wide structural and behavioral properties, which are illustrated in Fig. 3.

Despite the hierarchical relationships in the logical and cause-effect structures, all SEAs should be considered as one-type accurate within specific features. This means that complex activities are described by a set of the same type of model elements—SEAs. The SEAs describe the integration of activities, the formation of complex activities from their elements, which are, in fact, the organizational and management aspects of the CA. The resulting activity is directly modeled by specific elementary operations that are considered as constituent parts of the corresponding SEAs.

During the life cycles of SEAs, various resources can be used to implement technologies, organize actors, and form subject matters of CA. However, an important system-wide resource, used by all SEAs without exception, is the CA information model, which contains descriptions of SEA technologies, operational information, and other information objects specifying CA.

Based on the results of consideration of possible links between SEAs, it can be concluded that, at the system-wide level, the commonality of the links between the

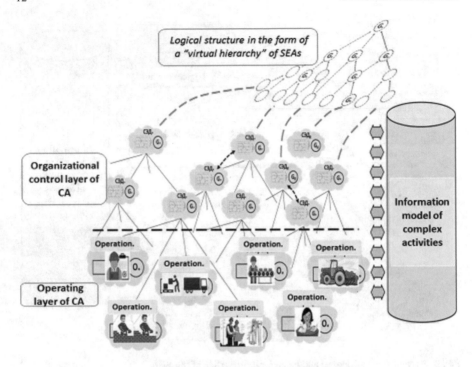

Fig. 4 Scheme for the implementation of a set of elements of complex activities

modeling elements have the nature of the exchange of information messages and/or the exchange of information through a common resource, i.e. the CA information model. The hierarchy and fractality of the logical structure are manifested when new elements of the CA are generated, but further, in the process of performing the CA, the hierarchy is realized through information exchange. The informational nature of the links allows us to talk about the certain autonomy of SEAs and about the "soft" links between them, the "disappearance of the hierarchy of SEAs", its transformation into an "implied", "virtual" form. In the process of performing CA, the hierarchy manifests itself only in the fact, that the directed information exchange between the elements occurs only in the pairs determined by the logical structure of the CA.

The analysis allows to base the developed model on the formalism of multiagent systems and extend it with the properties necessary for a unified description of the process of implementing any sets of CA elements (see Fig. 4, where the organizational, control layer of the CA can be conditionally called the layer of business processes, and operational layer of CA can be called the technological layer).

The SEAs play the role of autonomous similar agents in this formalism structure. The change in the status of SEA agents occurs according to a single scheme for realizing the life cycle of a CA according to the SEA process model. The specifics of agents (i.e. peculiarities of specific activities) are given by the logical and causal structures of each SEA, as well as by their specific features.

The logical model also defines the goal setting of the aggregate of SEA agents as a whole. The logical structure takes the form of a "virtual hierarchy" in the course of the implementation of SEA agents that are linked and interacts through messaging mechanisms and normative, a priori, operational information about the CA using a common information resource—the information model.

The aggregate of SEA agents can develop over time: some agents can generate new agents, others can cease to exist, and new technologies for the functioning of agents can be generated. Generation and termination of the existence of agents occur according to the life cycles of the elements of action, described by the process models of the SEA. The SEA agents are influenced by uncertainty events, some of which are generated by the SEA agents, other events are generated by the external environment.

The aggregate of SEA agents reflects the system-wide part of the complex activity that "connects" (organizes) the elements of activity into a single whole, forms a proper complex activity. The specific implementation of the activity itself is described by the elementary operations that are part of the SEAs.

Some of the SEAs can be referred to as "core activities", the other two parts are related to the organization and management, and the last one can be referred to "supporting activities".

The main types of the "supporting activities" are the creation of new technologies for activities (which are carried out by the corresponding SEA agents, whose implementation does not differ from the other ones) and the provision of resources (which is performed by the SEA agents that are also equivalent to the rest ones).

5 Conclusion

An adequate metaphor for the SEA agent is a certain "automatic machine" that reads the "program" (technology in the form of a logical, causal and process model) from the "repository" (information model of the CA) and executes the "program" taking into account external conditions, including a number of events of uncertainty.

According to this technology, the "SEA-automation" can generate other "SEA-automations". At the end of the "program", the existence of the "SEA-automaton" ceases. Individual "SEA-automation" form "programs" for other "SEA-automation" (i.e. create CA technologies), others perform the functions of the organization of resources. The interaction of "SEA-automaton" occurs through the exchange of messages, as well as by reading/writing normative, a priori, operational information from/to the "repository".

A dynamically changing set of equally organized, but realizing different "programs" "SEA-automation" provides a general description of complex activities (primarily the organizational and managing "layer" of CA). The direct implementation of the activity is modeled by the specific elementary operations ("operations automation"), each of which is associated with the "SEA-automation".

Thus, Fig. 4 and the above considerations illustrate the following statement: any complex activity can be represented as an extended multi-agent model.

This statement allows us to describe any complex acts as an aggregate of elements organized and united by common goal-setting, logical and causal structures. They are the following elements:

- specific elementary operations (representing elementary activities);
- the only or several control elements of the activity that implement (constant, or multiple, or continuous) verifications of the occurrence of certain conditions and initiation of the relevant specific elements of action, and also establish links between the CA entity as a whole, resources, and actors of the subordinate specific elements.

The given uniform description of any arbitrarily complex CA will allow building and effectively using the models of the CA, adequate for the requirements of the digital economy.

References

1. The program "Digital Economy of the Russian Federation", approved by the Decree of the Government of the Russian Federation of July 28, 2017, № 1632-r. http://government.ru/medi a/files/9gFM4FHj4PsB79I5v7yLVuPgu4bvR7M0.pdf
2. Belov, M., Novikov, D.: Methodology of Complex Activity, 320 p. Moscow, Lenand (in Russian) (2018)
3. Baicchi, A.: Construction Learning as a Complex Adaptive System, 131 p. Springer International Publishing, Switzerland (2015)
4. Gaubinger, K., et al.: Innovation and Product Management: A Holistic and Practical Approach to Uncertainty Reduction, 327 p. Springer, Berlin (2015)
5. State Standard 34.003-90 Information Technology. Set of Standards for Automated Systems. Automated Systems. Terms and Definitions
6. The cost of the iPhone 5c and iPhone 5s is calculated. http://hi-tech.mail.ru/news/iphone-5c-5 s-cost.html. Accessed 12 Mar 2015
7. Dalkir, K.: Knowledge Management in Theory and Practice, 2nd edn., 485 p. MIT Press, Cambridge (2011)
8. CIM data PLM glossary. http://www.cimdata.com/en/resources/about-plm/cimdata-plm-glos sary. PLM. Accessed 30 Jan 2017
9. Sauza Bedolla, J., et al.: PLM in engineering education: a pilot study for insights on actual and future trends. Product lifecycle management and the industry of the future. In: 14th IFIP WG 5.1 International Conference, PLM 2017, pp. 277–284
10. Duhon, B.: It's all in our heads. Inform.**12**(8), 8–13 (1998)

Methodology and Technology of Control Systems Development

Vladimir Burkov, Alexander Shchepkin, Valery Irikov
and Viktor Kondratiev

Abstract The development of new information technologies aimed at increasing the efficiency of activities in various fields is one of the most urgent tasks today. The leading Russian experts in strategic management have developed information technologies to improve the efficiency of activities in virtually all spheres of social and economic systems. Such technologies include the technology of working out development management systems (DMS). The research methodology consists of elements of control theory, systems theory, and system analysis methods. The main results of the research related to three models. The first model is related to risk management based on qualitative assessments. The second model is related to the synergetic effect that occurs when a pair of projects is included in the development program. The third model is devoted to the formation of the program calendar plan by the criterion of minimizing lost profits. The proposed system is successfully applied to the development of territorial development management systems, enterprises, personnel competence level and others. In the DMS technology, several mathematical models and methods are used. Basically, there is the "cost-effect" method and financial models. The chapter attempts to supplement the technology of DMS with a number of information technologies on the basis of mathematical models, the inclusion of which in the DMS technology increases the effectiveness of its application.

V. Burkov (✉) · A. Shchepkin · V. Irikov
V. A. Trapeznikov Institute of Control Sciences, Russian Academy of Sciences,
Moscow 117342, Russian Federation
e-mail: irbur27@mail.ru

A. Shchepkin
e-mail: sch@ipu.ru

V. Irikov
e-mail: irikov41@mail.ru

V. Kondratiev
Moscow State Automobile and Road Technical University, Moscow
125319, Russian Federation
e-mail: k-051310@mail.ru

© Springer Nature Switzerland AG 2019
A. G. Kravets (ed.), *Big Data-driven World: Legislation Issues and Control Technologies*, Studies in Systems, Decision and Control 181,
https://doi.org/10.1007/978-3-030-01358-5_2

Keywords Development management technology · Complex estimation
"Cost-effect" method · Risks · Scheduling

1 Introduction

The growth of the Russian economy is largely determined by the effectiveness of developing and implementing development programs for industries, regions, and enterprises. The Institute of Control Sciences of the Russian Academy of Sciences, together with the leading experts in strategic management, has created a technology for working out development management systems (DMS). This technology is based on three pillars: program-target management, project management and a complex of effective management mechanisms developed in the theory of active systems. A lot of works are devoted to that issue, for example, a new one [1]. One of the variants of the main stages of the technology is given below [2].

1. Analysis of the environment, the formulation of the purpose and criteria of the degree of its achievement.
2. Analysis and assessment of the potential of alternative ways of achieving the goal.
3. Selection of priority directions (programs) of changes that provide the main contribution to the achievement of goals.
4. Allocation of limited resources among them, maximizing the degree of the achievement of goals.
5. Formulation of the principles of requirements (policies, "rules of the game") to the management system.
6. Specification of the key indicators that characterize the performance of the executing agencies, and the requirements for their values that ensure achievement of the ultimate goals.
7. Development of a set of organizational measures to ensure the timely and high-quality implementation of programs (including assessment of activities, motivation, training, etc.)
8. Establishment of a system for regular monitoring of program performance.
9. On-line monitoring of results and adjustments to calendar plans, priority areas and, possibly, goals.

Analysis of the main stages of the technology of working out DMS shows that currently, it does not effectively use information technologies based on the optimization models and methods. The chapter proposes an interconnected complex of such information technologies, the inclusion of which in the technology of working out DMS increases its effectiveness. This complex consists of six information technologies:

– formation of a development program based on a system of integrated assessment of the state of the program (current and planned);
– accounting for multi-purpose projects;

Table 1 An example of the convolution matrix

4	2	3	3	4
3	2	2	3	4
2	1	2	3	3
1	1	2	2	3
2/1	1	2	3	4

- accounting for interdependent projects;
- risk management;
- adjustment of the program with the view of the additional costs when excluding projects from the program;
- formation of calendar plans.

Comment. The noted technologies were considered separately in [3]. *In this chapter, an attempt is made to present them as a single interconnected complex.*

2 Formation of the Program Based on the Integrated Assessment System

The program, as a rule, consists of several directions. In accordance with the methodology of working out DMS in each direction, the development potential is formed, that is, there are many projects that give an effect in this direction. The effect for each direction of the program is assessed in qualitative scales. The most popular is a four-point scale: bad—1, satisfactory—2, good—3, excellent—4. The boundary levels of the effect are defined: A_1, A_2, A_3, A_4. If the effect of E in the direction is less than A_1, then this is a catastrophic state in this direction. If $A_2 \leq E < A_3$, then the estimate is bad. If $A_2 \leq E < A_3$, then the estimate is satisfactory. If $A_3 \leq E < A_4$, then the estimate is good. Finally, if $E \geq A_4$, then the estimate is excellent. Further, for each i direction, the minimum costs s_{ij} required to reach the estimates j is determined. For this, one of the main information technologies in the technology of working out DMS is applied—the "cost-effect" method. All projects of this direction are ordered according to efficiency and selected according to this ordering until the effect is equal to or greater than the corresponding boundary value. The resulting minimum cost table (s_{ij}) is used in the integrated assessment system to form a development program that allows achieving the targets with minimum costs. The complex estimation system provides a dichotomous tree containing $(m - 1)$ vertices, where m is the number of directions. Each vertex of the tree corresponds to a convolution matrix of dual-purpose indicators (or generalized targets) [4]. An example of the convolution matrix of two-purpose indices is shown in Table 1.

Further on the basis of the table (s_{ij}) and the system of the complex estimation, a development program is elaborated using the method of dichotomous programming.

Table 2 Data for the first direction

i	1	2	3	4	5
a_i	15	20	14	18	13
c_i	5	8	7	12	13
q_i	3	2.5	2	1.5	1.0

Table 3 Data for the second direction

i	1	2	3	4	5
a_i	20	21	24	25	28
c_i	5	6	8	10	14
q_i	4	3.5	3	2.5	2.0

Table 4 The problem solution

4	2;29	3;42	3;49	4;51
3	2;11	2;24	3;31	4;43
2	1;5	2;18	3;25	3;37
1	1;0	2;13	2;20	3;32
2／1	1	2	3	4

4	43
3	25
2	11
1	0

The method of dichotomous programming is well known [5]. Therefore, let's give an illustration of the technology by the example.

Example 1 The number of directions is 2. For the first direction, there are five development projects, the data of which are given below, where a_i is the effect, c_i is the cost, $q_i = a_i/c_i$ is efficiency (Table 2).

We take the boundary values of the scale as $A_{11} = 4, A_{12} = 25, A_{13} = 44, A_{14} = 60$. Let the initial state be $A_{10} = 7$ (bad). Then, for transition to a state with a score of 2, it is necessary to add the effect $A_{12} = A_{12} - A_{10} = 18$; to transit to a state with a score of 3, we need to add $A_{13} = A_{13} - A_{10} = 37$; and to go to the state with the estimate $4 - A_{14} = 53$. Applying the "cost-effect" method, we obtain $s_{12} = 13, s_{13} = 20, s_{14} = 32$ for the first direction. For the second direction, there are also five projects, the data of which are given in Table 3.

We take the boundary values as $A_{21} = 8, A_{22} = 30, A_{23} = 50, A_{24} = 100$. Let the initial state be $A_0 = 10$. We have $s_{22} = 5, s_{23} = 11, s_{24} = 29$. Let us take the matrix as a complex estimation system (Table 1). The solution is shown in Table 4.

The first number is a complex estimate, and the second one is the cost of achieving it with the chosen option. The essence of optimization is that of all cells with the same first number, a cell is chosen whose second number is minimal. The table on the right shows the values of the minimum costs required to obtain a value for the integrated assessment. The decision itself, that is, the composition of the projects included in the program, is determined by the backward approach, for example, if the goal is to reach an estimate of 3, then the minimum costs are 25, and the program includes projects 1, 2, 3 of the first direction and project 1 of the second direction.

Table 5 Solution of option 1 problem

4	2;19	3;19	3;31	4;43		4	29
3	2;5	2;5	3;17	4;29		3	12
2	1;0	2;0	3;12	3;24		2	0
1	1;0	2;0	2;12	3;24		1	0
2/1	1	2	3	4			

The application of the integrated assessment system allows you to quickly adjust the program with reduced funding. For example, let's say that the financing of programs has decreased by half and has become equal to 12.5. In this case, the estimate 3 is unattainable. The target setting is adjusted, that is, the new goal is to reach the score 2. All the projects of the first direction are excluded from the program and the second project of the second direction is included in addition.

Comment. As you know, the "cost-effect" method will give an approximate solution to the problem of minimizing costs. With a large number of projects, the error of the method is insignificant. However, with a small number of projects, the error can be significant. In this case, to obtain a table of minimum costs for each direction, the problem of "knapsack" is solved by the method of dichotomous programming [5].

3 Accounting for Multi-purpose Projects

Multipurpose projects are those projects that give effect in several directions at once. If the number of multipurpose projects is not large, then all variants of entering the program of multipurpose projects can be considered (the number of variants is $2q$, where q is the number of multipurpose projects). For each variant, the task described in point 2 is solved. Of all the options, the best one is chosen.

Example 2 Let project 2 of the first direction and project 4 of the second direction be the same project with costs $c_{12} + c_{24} = 18$. There are two options.

Option 1. The project is included in the program. We should correct the target settings, subtracting the effects of the multipurpose project from them.

This is what we have for the first direction

$\Delta_{12} = A_{12} - A_{10} = 18 - 20 = -2$, $\Delta_{13} = 37 - 20 = 17$, $\Delta_{14} = 53 - 20 = 33$

Let us calculate $s_{12} = 0$, $s_{13} = 12$, $s_{14} = 24$.

This is what we have for the second direction

$\Delta_{22} = 20 - 25 = -5$, $\Delta_{23} = 40 - 25 = 15$, $\Delta_{24} = 65$

Let us calculate $s_{22} = 0$, $s_{23} = 5$, $s_{24} = 19$.

Substituting these data into the matrix of complex estimation Table 1, we obtain the solution given in Table 5.

To achieve a score of 3, the required cost units are $12 + 18 = 30$.

Option 2. A multi-purpose project is not included in the program. The target settings are not changed.

Table 6 Solution of option 2 problem

4	2;33	3;45	3;57	4;70		4	48
3	2;11	2;23	3;35	4;48		3	29
2	1;5	2;17	3;29	3;42		2	11
1	1;0	2;12	2;24	3;42		1	0
2/1	1	2	3	4			

Let us calculate for the first direction $s_{12} = 12$, $s_{13} = 24$, $s_{14} = 37$.

Let us calculate for the second direction $s_{22} = 5$, $s_{23} = 11$, $s_{24} = 33$.

Substituting these data into the matrix of complex estimation, we obtain the solution given in Table 6.

To achieve a score of 3, the required cost units are 29.

Let us choose the second option.

Projects 1, 3, 4 of the first direction and project 1 of the second direction are included in the program.

If the number of multipurpose projects is large, then the method of enumerating all the options leads to a large number of calculations. In this case, the network programming method becomes more efficient [5]. The idea is that the costs of multipurpose projects are divided arbitrarily into several parts according to the number of directions in which this project gives effect. Further, the problem is solved as in the case of single-purpose projects. Let us consider some basic theorems of the theory of network programming.

Theorem 1 *The solution of the problem with single-purpose projects gives a lower estimate of the costs for the original problem. The definition of the cost sharing of multipurpose projects, so that the lower estimate was maximum, is called the generalized dual task.*

Theorem 2 *The generalized dual task is a convex programming problem. The lower estimate can be used for the branch and bound method.*

Let us continue with the previous example. We'll divide costs 18 of the multipurpose project into two parts: 8 for the first direction and 10 for the second direction. We have obtained the problem solved in Example 1 with costs of 25. The solution received is not permissible, since Project 2 of the first direction has entered the program, and Project 4 of the second direction has not been included. We will increase the costs of the project 2 by 4. To obtain the score 3 in the first direction, it is necessary to include the same projects 1, 2 and 3 with the costs 24 in the program. However, in the second direction, project 4 becomes the most priority and is included in the program with a cost of 5 units. The solution obtained is admissible and, therefore, optimal one.

Fig. 1 An example of a
graph of interdependencies

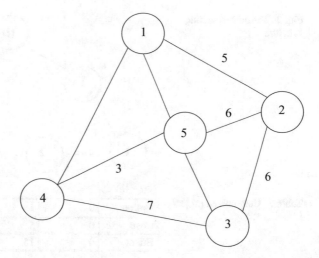

4 Accounting for Interdependent Projects

Projects, the inclusion of which in the program gives an effect more than the sum of
the effects of individual projects (the so-called synergistic effect), are called interdependent projects. In this case, the "cost-effect" method is not applicable. Let us
define the graph of interdependencies. The vertices of the graph correspond to the
projects. Two vertices i, j are connected by an edge if the corresponding projects are
interdependent. The length of the edge is equal to the additional effect b_{ij}. Figure 1
shows an example of a graph of interdependencies for the five first direction projects
of Example 1.

We'll describe the algorithm for solving the problem, consisting of three stages.

Stage 1. A pairing in a graph is a set of edges that do not have common vertices.
Let us remove a lot of vertices from the graph so that the remaining edges form a
combination of pairs. So, if the vertex is removed from the graph (in Fig. 1), then
the remaining edges (1, 2) and (3, 4) form a combination pair.

Stage 2. Suppose that the number of removed vertices is q. We'll consider all $2q$
variants of including the removed projects in the program. If in the case under consideration the removed vertex i is included in the program, then for the adjacent (not
removed) vertices j a synergetic effect is added to their effects. The total effect for
the removed vertices is calculated taking into account their interdependence. The
target settings Δ_2, Δ_3, Δ_4 are adjusted.

Stage 3. For the remaining (not removed) projects, the task is to achieve the targets
with minimal costs. The problem is solved by the method of dichotomous programming [5]. The structure of the dichotomous representation is chosen in such a way
that the vertices connected by an edge are on the lower level of the dichotomous tree.
The corresponding structure for the graph of Fig. 1 is shown in Fig. 2.

Fig. 2 The corresponding structure

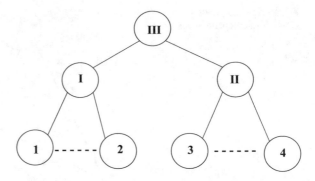

Table 7 Generalized project I

Option	0	1	2	3
Costs	0	5	8	13
Effect	0	15	20	40

Table 8 Generalized project II

Option	0	1	2	3
Costs	0	7	12	19
Effect	0	14	18	39

Table 9 Solution of option 1 problem

3	19;39	24;54	27;59	32;79
2	12;18	17;33	20;38	25;58
1	7;14	12;29	15;34	20;54
0	0	5;15	8;20	13;40
II / I	0	1	2	3

Example 3 Let us consider the solution of the problem by the method of dichotomous programming. Since the number of removed vertices is 1, there are two options.

Option 1. Project 5 was not included in the program.

1st step. We'll consider interdependent projects 1 and 2. The solutions are presented in Table 7.

2nd step. Let us consider interdependent projects 3 and 4. The solution is presented in Table 8.

3rd step. Let us consider generalized projects I and II. The solution is presented in Table 9.

Since $\Delta_{12} = 18$, $\Delta_{13} = 37$, $\Delta_{14} = 53$, then $s_{12} = 8$, $s_{13} = 13$, $s_{14} = 20$.

Option 2. Project 5 is included in the program. Let us adjust target settings $\Delta_{12} = 18 - 13 = 5$, $\Delta_{13} = 37 - 13 = 24$, $\Delta_{14} = 53 - 13 = 40$.

Let us add the appropriate synergistic effects to the effects of projects $a_0 = 15 + 4 = 19$, $a_1 = 20 + 6 = 26$, $a_2 = 14 + 2 = 16$, $a_3 = 18 + 3 = 21$.

Table 10 Solution of option 2 problem

3	19;44	24;63	27;70	32;94
2	12;21	17;40	20;47	25;71
1	7;16	12;35	15;42	20;66
0	0	5;19	8;26	13;50
II / I	0	1	2	3

Table 11 Low-risk projects' "cost-effect" dependence

Option	0	1	2
Costs	0	12	25
Effect	0	18	31

We'll solve the problem by the method of dichotomous programming. The table of the third step is given in Table 10.

Let us calculate $s_{12} = 5 + 13 = 18$, $s_{13} = 8 + 13 = 21$, $s_{14} = 13 + 13 = 26$. Comparing the two options, we'll choose the first one.

5 Risk Management

Risk management includes such basic processes as identification of risks, determination of their main characteristics, and choice of ways of responding to risks (reduction, transfer, evasion, and acceptance). The risk is described by two indicators–the likelihood of a risky event and the damage in its occurrence. A general characteristic is the degree of influence (or rank) of the risk, which is defined as the expected damage. Recently, a lot of attention [6, 7] is drawn to the study of risk management tasks based on qualitative estimates of their characteristics. The point is that in practice, qualitative risk assessments are most commonly used. The simplest is a two-evaluation scale (low probability, high probability, small damage, large damage, a low degree of influence, a high degree of influence). This is understandable, since the project is, by definition, a unique event, which does not allow you to fully rely on statistical data. One of the ways to reduce the risk of the program is to limit the financing of high-risk projects. We'll describe the modification of the "cost-effect" method, with the account of the availability of high-risk projects. We are going to consider this method based on the data of Example 1 (first direction). Let projects 1, 2 and 3 be high-risk ones. Let's take a restriction on financing high-risk projects $R_\theta = 12$.

1st step. Let's build the "cost-effect" dependence for the low-risk projects (Table 11).

2nd step. Let's build the "cost-effect" dependence for the high-risk projects (Table 12).

3rd step. Let's form a final table (Table 13).

Table 12 High-risk projects "cost-effect" dependence

Option	0	1	2	3
Costs	0	5	8	12
Effect	0	15	20	29

Table 13 Final solution data

2	25;31	30;46	33;51	37;60
1	12;18	17;33	20;38	24;47
0	0	5;15	8;20	12;29
	0	1	2	3

Let's calculate for target settings $\Delta_{12} = 18$, $\Delta_{13} = 37$, $\Delta_{14} = 53$.
$s_{12} = 8$, $s_{13} = 20$, $s_{14} = 37$.
To achieve assessments of 2 and 3, a high-risk project 2 is included in the program, to achieve an assessment of 4, high-risk projects 2 and 4 are included in the program.

6 Adjusting the Composition of the Program

The tasks of operational management associated with adjusting the composition of the program arise for many reasons. First, in the case of a reduction in the amount of funding for the program. Secondly, in the case of the emergence of new high-performance projects, reducing the effects of projects included in the program, increasing the risk projects, etc. The peculiarity of the tasks of adjusting of the composition of programs is the fact that when the project is excluded from the program, additional costs arise, related to the breaking of contracts, compensation payments to performers, etc. With a large amount of these costs, it is more profitable to leave the project in the program than to exclude it. Let's make the following denotations: additional costs when excluding project i from the program, Q—a lot of new projects, P—a lot of old projects.

The cost constraint will have the form

$$\sum_{i \in Q} c_i x_i + \sum_{i \in P} (c_i x_i + l_i(1 - x_i)) \leq R$$

or

$$\sum_{i \in Q} c_i x_i + \sum_{i \in P} (c_i - l_i)x_i \leq R - \sum_{i \in P} l_i$$

Table 14 The duration of the projects and priorities

i	1	2	3
τ	15	10	5
p	1	2	2.8

Example 4 We'll take the project data from Example 1 (direction 1). Let projects 1, 2, 3 be new, and projects 4, 5 be old ones. Let's consider $l_4 = 7$, $l_5 = 6$. The ordering of projects by efficiency has the form

$$4 \to 1 \to 2 \to 3 \to 5$$

Let $R = 31$. Applying the "cost-effect" method, we get that the program includes projects 4, 1 and 2, that is, project 4 remains in the program, although there is a more efficient project 3.

7 Formation of Calendar Plans

When the composition of the program's projects is formed, it is necessary to develop a calendar plan for the implementation of the program with a specified funding schedule. As part of the criterion is the value of the loss of profit $= \sum_i a_i t_i$, where t_i is the completion time of the *i-th* project. The task is a complex optimization problem that does not have effective precise decision algorithms.

Consider the heuristic algorithm, which is based on the priority rules of projects. Let's single out two priority rules.

The first rule is the above rule $q_i = a_i/c_i$, that is, the project's effectiveness in terms of cost. The second rule $p_i = a_i/\tau_i$, τ_i—the duration of the *i-th* project, characterizes the effectiveness of the project in time. The first rule gives a solution close to optimal for the case when the program is executed and financed by periods, and each project can be fully executed during a given period [8]. The second rule gives an optimal solution in the case when projects are executed sequentially. In the general case, we'll take a convex linear combination of these rules

$$r_i(\alpha) = \alpha q_i + (1-\alpha)p_i, \quad \text{where } 0 \leq \alpha \leq 1$$

Solving the problem for different values of α, we choose the best solution.

Example 5 As a result of solving the task of forming the program in Example 1, projects 1, 2 and 3 for the first direction were selected. Let the integral schedule of financing these projects be given. The Table 14 shows the duration of the projects τ_i and the priorities p_i.

The term of the program completion is T = 20.

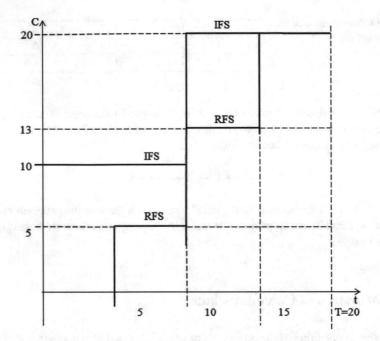

Fig. 3 An integral financing schedule

Let the integral financing schedule (IFS) look like in Fig. 3. In order to verify the financial feasibility of the program, we will construct an integral financing schedule for the start of projects at the latest possible dates [a right-shifted financing schedule (RFS)] (Fig. 3). Since the RFS is not higher than the IFS, the program is realizable.

Let's consider three variants.

1. $\alpha = 0$. The first project to be executed by priority is the project 3. Projects 1 and 2 begin only at the time $t = 10$ (projects do not start until all the means required for their fulfillment have arrived). Loss of profits will *be* $F = 14 \cdot 5 + 20 \cdot 35 = 770$
2. $\alpha = 1$. The first project to be executed by priority is the project 1. The lost profit will be $F = 15 \cdot 15 + 20 \cdot 34 = 905$
3. $\alpha = \frac{1}{2}$. The highest priority is again project 3. Loss of profit is 770.

The best option is to fulfill project 3 first.

8 Conclusion

In our view, the set of models and methods proposed should be developed in various directions. Thus, it is desirable to supplement the system of integrated assessment with methods of forming matrix convolutions, with the account of the preferences of managers. Risk management models (for example, risk mitigation models, models and methods of scheduling, etc.) should further be developed. An important task is to apply the proposed information technologies in the development of software products in various fields, such as described in [9, 10], for example, and furthermore.

References

1. Lamperti, G., Zanella, M., Zhao, X.: Introduction to Diagnosis of Active Systems, 353 p. Springer International Publishing, Switzerland (2018)
2. Ansoff, H.: Strategic Management, 251 p. Palgrave Macmillan, UK (2007)
3. Al Khouri, A.: Program Management of Technology Endeavours: Lateral Thinking in Large Scale Government Program Management, 298 p. Palgrave Macmillan, UK (2015)
4. Reichwald, R., Wigan, R.T.: Information, Organization and Management, 536 p. Springer, Berlin (2008)
5. Burkova, I.V.: The method of network programming. Problems of nonlinear optimization. Autom. Telemechanics (11) (2009)
6. Kravets, A., Kozunova, S.: The risk management model of design department's PDM information system. Commun. Comput. Inform. Sci. **754**, 490–500 (2017)
7. Shcherbakov, M., Groumpos, P.P., Kravets, A.: A method and IR4I index indicating the readiness of business processes for data science solutions. Commun. Comput. Inf. Sci. **754**, 21–34 (2017)
8. Szostak, R.: The Causes of Economic Growth: Interdisciplinary Perspectives, 373 p. Springer, Berlin (2009)
9. Goroshko, I.V., Smirnov, M.V., Bondarenko, J.V.: The task of coordinated interaction in the international police cooperation. In: IEEE Proceedings of 10th International Conference Management of Large-Scale System Development, MLSD. https://doi.org/10.1109/MLSD.2017.8 109627 (2017)
10. Bondarenko, Y.V., Azarnova, T.V., Kashirina, I.L., Goroshko, I.V.: Mathematical models and methods of assisting state subsidy distribution at the regional level. In: International Conference "Applied Mathematics, Computational Science and Mechanics: Current Problems". Voronezh, Russian Federation. 18–20 Dec 2017. https://doi.org/10.1088/1742-6596/973/1/012061

Principles of Mathematical Models Constructing Based on the Text or Qualitative Data of Social Systems

Alexey Lebedev, Andrey Shmonin, Fyodor Vasiliev and Vadim Korobko

Abstract The predominance of textual or qualitative data in the information resource of the social system leads to difficulties in formalizing the process of preparing and making managerial decisions based on such types of data. The research is based on the theory of social management, the analysis of scientific literature on the topic of research, as well as on the theory of similarity. A set of models providing intellectual analysis of textual and qualitative data is offered. An overview of models of this kind is given in this chapter. Their advantages and disadvantages are noted for processing various types of data available in social systems. The models of such social systems are considered for the applicability of the model designed for one social system on the basis of quantitative data to regulate other systems on the basis of qualitative data. It is shown that most of the processes occurring in social systems are non-linear in nature. The chapter determines the possibility of more complete use of textual or qualitative data of a social system for modeling its activities on the basis of similarity theory. The obtained results create prerequisites for the development of methods for models scaling and methods of transferring them to similar systems.

Keywords Social system · Qualitative data · Model verification
Similarity theory · Model scaling

A. Lebedev (✉) · A. Shmonin · F. Vasiliev
Academy of Management of the Ministry of Internal Affairs of Russia,
Moscow 125993, Russian Federation
e-mail: lebedevavic@rambler.ru

A. Shmonin
e-mail: shmonin@hovrino.net

F. Vasiliev
e-mail: vasilev17@mail.ru

V. Korobko
All-Russian Public Organization «Business Russia», Moscow, Russian Federation
e-mail: vkorobko@mail.ru

© Springer Nature Switzerland AG 2019
A. G. Kravets (ed.), *Big Data-driven World: Legislation Issues and Control Technologies*, Studies in Systems, Decision and Control 181,
https://doi.org/10.1007/978-3-030-01358-5_3

1 Introduction

At present, information technologies are actively used for the support of decision-making in social systems. However, this situation is typical only for data presented in a numerical format. Along with quantitative data, a significant amount of text data is circulating in the social system. Exact estimates of their volumes do not seem to exist. The following example can illustrate the scale of the problem: the amount of data of the electronic document management system functioning for 2 years in one of the federal executive bodies has exceeded the volume of statistical data accumulated there for the period of almost 20 years. At the same time, a huge array of unencrypted text documents is not taken into consideration. The vast majority of documents circulating in the social system (orders, job descriptions, administrative regulations, memoranda, references, etc.) are presented in the form of text documents.

We assume the following logical statement as a canonical form of a problem: "V is given, W is required" (or in the form of a short entry <V, W>). The specified record includes the set of V states of the object in which it is at the present time and the set of W states of the object to which it is necessary to transform (the required state of the object) [1]. This required state, when solving the problem for the first time (and in most cases), is a verbal description formulated by the decision-maker. In a short time (1–3 days), by abstraction, it can only be translated into a qualitative description (2–3 possible situations). The solution of the task is a set of operators, applying which in a certain order (algorithm), we move from the current state to the desired one. A situation is possible when such operators (operator) transferring a set from one state to another do not currently exist, or they have not yet been formed and require a certain amount of time to form them.

In general, when making managerial decisions in a social system, in most cases, the "input and output" of a problem is presented in the form of text documents (or their elements), and its solution (ways, methods of data processing) remains behind the scenes and is built on the employee's experience, the decision maker, and his intuition. Thus, the task of making managerial decisions is not formalized, is not based on quantitative data, and from the point of view of applied mathematics, is not subject to automation.

When working with quantitative data coming from the information sources of social systems, they are used to construct appropriate hypotheses (theories) that explain the behavior of a social system in a certain period of time or under certain circumstances, which in turn provides an understanding of the situation.

A complete theory of any workflow can only be built on sufficient (or, better, redundant) data, which contain many variables and conditions that affect them. Initially, the theory is also constructed in text form, although already in its formation, mathematical objects and structures can be used. Mathematical models are tools that help in the creation of such a theory (confirmation of a hypothesis). In other words, models are theories having mathematical formalization.

The popularity of mathematical models of social systems began to increase in the 1950s and has increased significantly since the 1980s, including because of the use of personal computers and the possibility of accumulating and processing large amounts of data (primarily quantitative ones). Each mathematical model, as a rule, gives a significant effect in its field, and sometimes, goes far beyond its limits.

2 Principles of Construction and Types of Mathematical Models

To understand the importance of mathematical modeling, it is useful to reconsider the solution of a non-formalized problem of making managerial decisions.

A text message, which initiates the process of making managerial decisions, is introduced into the social system. If the specialist understands that the management decision somehow depends on external or internal conditions of the workflow, then he tries to collect the necessary data and analyze them. Even if the data is received, their overall assessment represents some levels of the independent variable (for example, such as low, medium, high). The hypothesis arising from the employee is usually expressed in an oral form (for example, "it will be worse in case A than in case B"), so we call it the verbal model.

Further, it is necessary to verify this hypothesis, at least for the absence of a difference between cases A and B. To test this hypothesis, we need at least statistical data on the state of the problem, which the specialist, as a rule, does not possess. In this regard, he makes a subjective decision. The experience of the authors of the chapter shows that at the lower level of the social system (the level of subdivisions), 90 and more percent of decisions are taken in a similar way.

Suppose that a specialist still has the necessary data and skills for statistical analysis (which is more an exception than a rule). In this case, a statistically significant difference allows him to draw a conclusion about the advantage of the case B. But if the difference is not statistically significant, then he returns to the initial state of management decision making, when uncertainty prevails in the data and, in fact, the decision cannot be rendered, as it is not justified. In this regard, the specialist again makes a subjective decision.

It becomes obvious that it is necessary to create such models that should provide intellectual analysis of text and qualitative data and explain the behavior of the system (workflow), at least at the level of abstraction, which allows to build data in the ordinal scale with an insignificant number of gradations (3–5).

Existing mathematical models based on quantitative data have nothing to do with verbal models or data with a qualitative difference between them. To draw the conclusion about the advisability of choosing one of the situations (A or B), that is, to confirm the difference between the situations, we must go beyond the level of specification available in the verbal models [2].

In a mathematical model, hypotheses are expressed in the form of mathematical equations, computer algorithms, or other modeling procedures. Accordingly, mathematical models go beyond the scope of qualitative predictions, such as "the value

of the parameter in state A will be higher than the value of the parameter in state B".
The mathematical model, in this case, will allow us to formulate exact quantitative
predictions, such as "the value of the parameter in state A will be 20% higher than
the value of the parameter in state B", which can be verified experimentally.

In addition, the use of mathematical models allows us to move from linear and
equilibrium models to models with nonlinear relations and dynamic processes that
can accurately reflect the complexity of the processes taking place in the social
system.

Suppose, we have data presented on the ordinal scale with an insignificant number
of gradations (3–5). To solve the problem of explaining the behavior of the system
(workflow), at least at the abstraction level, it is possible to propose an axiomatic
method of mathematical modeling that involves the replacement of the process itself
with a set of simple positions or axioms designed in such a way that the observed
behavior of the social system can be deduced logically of them. Each axiom itself
is a fundamental assumption about the process and takes the form of a declarative
restriction or existence statement, for example "the parameter value is always greater
than zero" or "the value of the variable y exists greater than zero, so the effect on
the system does not depend on the change in the incoming quantity from the level
A to the level of $A + y$". Together, this set of declarative constraints and rules allows
you to impose constraints on variables that will be sufficient to uniquely identify the
model and available data.

For example, the expected utility model proposed by Chater [3] can be chosen as
such a model. It studies the choice between risk situations, with only one and several
possible outcomes. Suppose, that n such outcomes are analyzed. If we consider the
set of states of the V object in which it is at the present time (by qualitative values),
designate the outcome vectors through w_i, and the probabilities corresponding to
each of them by p_i, then the expected utility can be represented as the expression:

$$F(w_i, p_i) \to \max,$$

where F is the multiplicative combination of probabilities and outcomes after some
transformations.

Thus, in this case, we can talk about the need to specify only a mathematical
transformation of F, while the initial data can be completely specified in text form
with their transformation into qualitative data, declarative constraints and a lot of
possible alternatives. We should note that probabilities can also be set by the method
of expert assessments (which in most cases is realized in practice).

There are many varieties of models of this kind, which differ in the way of utility
measurement, the permissible types of probability conversion of F, and the method
for measuring the outcomes of w_i. It is obvious that we are only interested in the case
when the "inputs and outputs" of the problem are presented in the form of qualitative
data.

The model predicts that the decision-maker will always choose alternatives with a
higher expected performance, which is quite typical for the economic social systems,
but not for the systems created within the framework of public tasks, where the

main alternative will be to minimize the risk in the situation of making managerial decisions.

It is evident, that the mathematical model is fully workable for qualitative data, but the creation of a universal utility function, that completely determines the recommendations for the decision-maker regarding all possible options, is an extremely difficult task. In this case, the testing of several different functions of expected utility should be recommended, so that to select the most suitable of them.

In addition, as shown in [4], an additional declarative rule is possible for this model: for any choice of situation B, such that the choice of situation A is preferable to the choice of situation B, which in turn is preferable to the choice of situation C, there is a single-valued probability q, such that there is an additional situation of indifference between the choice of the situation B and the choice of the composed of q chance probability situation A and $(1 - q)$ of the chance probability of the situation C in which A is chosen with probability q and C is chosen with probability $(1 - q)$.

The described axiomatic method is an approach to mathematical modeling, which requires a consistent study of a number of managerial decisions. Complexities, in this case, relate to the construction of axioms for managerial decisions and their consistent refinement. All assumptions of the model must be explicitly stated in verifiable axioms. Axiomatic models are quite understandable in the sense that they generate management decisions. Moreover, because of the logical rigor of their design, axiomatic models are durable. For these reasons, many scientists believe that the knowledge gained from axiomatic modeling is of the highest quality [5].

Another approach to the verbal description of the data may be an algebraic model that is essentially a generalization of the standard linear regression model in the sense that it accurately describes how the original model parameters are combined to obtain an algebraic equation. Algebraic models are usually easy to understand because there is an essential connection between the descriptive (verbal) theory and its mathematical representation.

The simplest example of algebraic models is the general linear model, which is represented by linear combinations of input parameters in the form of $y = ax + b$, in which parameters (a, b) measure the degree of their effect on the output value. These models are simply built, but in the authors' opinion, their creation should be based on a criterion that allows you to select the factors that make up the model. As a selection criterion, it is possible to choose the minimization of the partial pair correlations of the factors included in the model.

The model [6] constructed in the light of the proposed criterion, which included the following factors:

– the ratio of the monetary allowance of the staff of the Department of Corrections to the average accrued wages in the country—x_1;
– the number of the employees of the Department of Corrections who were dismissed during the reporting period in relation to the number of the acting personnel—x_2,
– load per one employee—x_3;

has the form:

$$y = -0.27 \cdot x_1 - 0.96 \cdot x_2 + 0.05 \cdot x_3 + 0.2$$

A dependent value (y) is the indicator "employees convicted in the reporting period per 1000 people".

This model is rather coarse (the determinism coefficient is 0.61, while it has significant standard errors of coefficients and constant), but for all its inaccuracies, it finds a sufficiently good qualitative confirmation of the behavior of the model variable from the dependent factors.

Models of this kind are very good for similar [7] social systems, when proving their applicability in one social system (its subdivision) based on quantitative data; they are used for the regulation of other systems based on qualitative data (regulation of the controlled parameters of the model in certain intervals by increasing or decreasing them).

In this case, quantitative data are required for the construction of the model itself, but its use is also possible for qualitative data. Building a model for such social systems is possible using the data of only one or several units, which significantly reduces the cost of their collection and processing.

The algorithmic model of decision-making based on qualitative data can be defined in terms of the procedure for processing the order of interaction of internal workflows in the social system. The associated processes are often so complex and interconnected that the model predictions cannot be obtained by constructing a simple algebraic equation (the problem of multicollinearity or the interdependence of incoming task parameters). In this case, to formulate the output based on model representations, it is required to simulate the dynamic processes by means of a computer with the help of the random number generator [8]. This requires only a qualitative relationship between the individual "inputs and outputs of the system". The process is started as a sequence of random effects that are intended to represent the relevant activities for making managerial decisions. As a result, an output value is generated, which usually corresponds to the decision or action taken by the decision-maker. The construction of the algorithmic model is based on the fact that the simulation of workflows accurately reproduces the actions of a particular specialist. In contrast to the axiomatic modeling approach, in which each assumption is theoretically justified, the algorithmic model system makes many assumptions about workflows related to the behavior of a particular specialist that cannot be empirically confirmed, since they are not directly observable.

This model provides significant opportunities for adjusting its internal structure and conducting rapid observations of the behavior of the social system. One of the advantages of this approach is that it allows you to work with ideas that cannot yet be expressed in precise mathematical form. In addition, it is possible to expand the model by using complex cognitive and neural processes based on the verification of existing processes and the above considerations [9].

The main disadvantage of algorithmic modeling is the lack of connection between parts of the model and their mental analogs. The advantage is flexibility, which allows you to quickly create and test models of this kind. To minimize this problem, algorithmic models must be developed with minimal assumptions, and, in addition, care should be taken to ensure that all assumptions are well-founded and plausible.

Development of algorithmic models can be represented by cognitive models, which are described by multilayered networks of interconnected nodes. At the same time, data can also be (and are) of a qualitative nature [10]. Model predictions are generated by encoding the activation of a set of "input nodes", which then transmit activation through a series of "hidden nodes" and transform the source module into new codes or functions until the actual activation reaches the "output node", which is a solution to the management task.

Models of "connection" can be described as a separate subclass of algorithmic models. The key difference is that connectivity models are focused on learning the patterns in data through training. In fact, the model learns to create the right data pattern, adapting its experience with input data, strengthening and loosening the links, similar to the processes which occur in the human brain in the cause of learning.

This flexibility allows you to model very complex data sets. In fact, it was proved that a model with a sufficiently large number of hidden units can approximate any continuous non-linear input-output relationship with the required degree of accuracy [10]. The use of these models is very promising for social systems, since, in the authors' opinion, most of the processes occurring in them are of a non-linear nature. At least, the construction of linear algebraic models based on the available data in most cases gives unsatisfactory results precisely for a qualitative agreement with the available data.

Just as verbal models are constructed on the basis of interpreting the available more complete and accurate data and some intuitive assumptions about the workflows taking place in the social system, mathematical models are built on an even higher level of assumptions, and, naturally, in the first approximation, they can give either a wide range of solutions, or unacceptable solutions. Nevertheless, the process of working with mathematical models is very interesting precisely in terms of rethinking existing assumptions, points of view on the problem and interpretation of the available data.

Unlike predictions of verbal models that are qualitative in nature and expressed in the oral form, the conclusions drawn by mathematical models characterize quantitative relationships that clearly determine the influence of one variable on another.

In this regard, the authors note that mathematical models are built on well-defined mathematical objects. For example, the set used to specify a function is specified either by a rule or an enumeration. For social systems, a situation is typical where even the initial set of objects is given only by enumeration and is dynamic and heterogeneous at the same time.

The model can be defined as a parametrized family of probability distributions $M = \{f(y|w), w \in W\}$, where $y = (y_1, \ldots, y_n)$ is a data vector for n observations, w is a parameter specifying parameters of a model; and $f(y|w)$ are is probability density functions that determine the probability of observing y for a given w; and, finally, W is the parameter space [5].

In this regard, for models built on quantitative data, it is practically impossible to establish the need for a model (only sufficiency), since someone can almost always come up with a model based on other assumptions. But at the same time, the new model can also reflect the available data well.

For models built on qualitative data, you can come up with an even larger number of models equally well reflecting the available data. However, due to the greater simplicity in their construction and analysis, the costs of constructing mathematical models on qualitative and quantitative data will not be comparable. At the same time, we still forget about the issue of data collection in the social system.

Since the data in the social system are formed on the basis of the reflection of the situation by a particular employee, they are a priori more expensive than such data in technical systems and less accurate.

Models based on qualitative data possess certain advantages: they are simpler, faster reconfigurable, almost not subject to noise and disturbing effects. Considering all the above factors, as well as the unreliability and inaccuracy of the available data, which in most cases can only be interpreted qualitatively, it seems that there are simply no alternatives to models built on qualitative data in social systems.

A simple model, as a rule, has several parameters and allows you to make clear and easily understood conclusions, which is especially appreciated by the decision-maker. On the other hand, the complex model will have much more parameters, which will make it more flexible and capable of predicting many different data sets with high accuracy, by fine-tuning the parameters.

However, a very complex model works, as a rule, in a very narrow range of data, since its parameters can be tuned to fit almost any data array, including random data. As such, a complex model can often provide redundancy by using random noise that is characteristic of a particular data set, but it does not necessarily result from some dependency underlying the data.

An important goal of modeling is to identify hypotheses that generate accurate data, so the goal of modeling is to select the model that can best generalize and not the one that best fits one data set. This is another plus of models built on qualitative data.

The main idea proposed by the authors is to divide the data available to the social system into two additional subsets. One subset, called a training or calibration set, is used for modeling by constructing descriptive models on qualitative data. Another subset, called a set of settings, is considered as a "future" data set and is used to check estimates from a set of mathematical models built on quantitative data, but taking into account all the axioms and rules prescribed for the first models.

If the mathematical model turns out to be inappropriate for certain sets, it makes sense to work out new axioms and rules for these data and try to restructure the mathematical model.

We should note that mathematical models based on qualitative data can be useful precisely in terms of restructuring existing axioms. This utility was identified by S. Karlin, who said: "The purpose of models is not to match data, but to sharpen questions" [11].

3 Conclusion

Mathematical modeling involves the transformation of ideas and assumptions into mathematical abstraction. The advantage of mathematical models built on qualitative data is manifested in the speed of their development and simplicity.

The model is as valuable as the hypotheses generated by it are true. This means that mathematical models do not end with their formulation in the form of a set of rules or a mathematical equation, but rather are the bricks on the path of scientific knowledge. Newly created models are always used to expand the knowledge gained from the previous models.

Modeling has opened up new ways of solving problems, but at the moment it does not have much success in social systems, which, in the authors' opinion, is not only related to the specifics of making managerial decisions in social systems, but also to the ways and means of presenting data in them. The success of mathematical models is evident on the assumption that the study of interactions between correlation variables provides an understanding of the workflows in the social system. Unfortunately, this is not always feasible in social systems, because of the fact that they have a long-established and complicated structure, and, moreover, inherit historical mechanisms for making managerial decisions.

Practical modeling is always tied to a specific area, be it a phenomenon, task, workflow or a social system [12, 13]. Having achieved success in their field, developers are faced with the problem of expanding the scope of their models to explain the performance of other tasks, additional phenomena or to combine several levels of description. The spread of the model to other fields is associated with serious risks of its complication. The development of methods for models scaling and methods of transferring them to similar systems will be an important step in the modeling of social systems.

References

1. Lebedev, A.V.: Mathematical model of crime in conditions of real control. Bull. Vladimir Law Univ., Federal Penitentiary Service of Russia **4**(21), 92–99 (2004)
2. Eom, S.: The intellectual structure of decision support systems research. In: Decision Support: An Examination of the DSS Discipline. Annals of Information Systems, vol. 14, pp. 49–68. Springer, New York (2011)
3. Chater, N., Tenenbaum, J., Yuille, A.: Probabilistic models of cognition: conceptual foundations. Trends Cogn. Sci. **10**, 278–291 (2006)
4. Barbera, S., Hammond, P., Seidl, C. (ed.): Handbook of Utility Theory: Volume 2 Extensions, 626 p. Springer, US (2004)
5. Luce, R.D.: Utility of Gains and Losses: Measurement Theoretical and Experimental Approaches. Lawrence Erlbaum, NJ (2000)
6. Lebedev, A.V., Babkin, A.A., Shakhov, O.A.: On the use of the mathematical model with information support for decision-making in the Federal Service of Corrections of Russia. Vestnik of the Vologda University of Law and Economics of the Federal Service of Corrections of Russia **2**(14), 73–80 (2011)

7. Zezula, P., et al.: Similarity Search: The Metric Space Approach, 220 p. Springer, US (2006)
8. Pitt Francis, J., Whiteley, J.: Guide to Scientific Computing in C++: Undergraduate Topics in Computer Science, 287 p. Springer International Publishing, Heidelberg (2017)
9. Busmeyer, J.R., Diederich, A.: Cognitive Modeling. SAGE Publications, Inc., Thousand Oaks (2010)
10. Krawczak, M.: Multilayer Neural Networks: A Generalized Net Perspective. Studies in Computational Intelligence, 182 p. Springer International Publishing, Heidelberg (2013)
11. Karlin, S.: The 11th R.A. Fisher Memorial Lecture given at the Royal Society 20th Meeting in April (1983)
12. Goroshko, I.V., Smirnov, M.V., Bondarenko, J.V.: The task of coordinated interaction in the international police cooperation. In: Proceedings of 10th International Conference Management of Large-Scale System Development, MLSD, IEEE (2017) https://doi.org/10.1109/mlsd.2017.8109627
13. Bondarenko, Y.V., Azarnova, T.V., Kashirina, I.L., Goroshko, I.V.: Mathematical models and methods of assisting state subsidy distribution at the regional level. In: International Conference on Applied Mathematics, Computational Science and Mechanics: Current Problems, 18–20 Dec 2017, Voronezh Russian Federation. https://doi.org/10.1088/1742-6596/973/1/012061

On the Possibility of an Event Prediction with Limited Initial Statistical Data

Alexander Betskov, Valery Makarov, Tatiana Kilmashkina
and Anatoly Ovchinsky

Abstract In the technogenic sphere, and not only, there are socially resonant events that bring significant damage to the social sphere. An adequate prediction of such events is required to prevent and minimize damage from them. There are various methods for predicting accidents and disasters, determining the ultimate level of probability of occurrence or the absence of an event. However, forecasting, as a rule, is complicated by the small number of them. The general statistical aggregate of risk events is limited and does not allow us to apply the theory of probability. As a result, a new method of mathematical statistics has been developed, the application of which makes it possible to predict events with a certain probability on the basis of relatively small statistical data. It is proposed that a new approach to determine the probability of interesting events, with a limited general statistical sample, will allow to predict possible threats, with the greatest likelihood.

Keywords Probability theory · Probable threats · Limited sampling
Limited statistics · Theory of resampling · General statistical aggregate
Forecasting

A. Betskov · V. Makarov (✉) · T. Kilmashkina
Academy of Management of the Ministry of Internal Affairs of Russia,
Moscow 125993, Russian Federation
e-mail: ovorta@mail.ru

A. Betskov
e-mail: amvd-6@bk.ru

T. Kilmashkina
e-mail: kilmashkinnf@yandex.ru

A. Ovchinsky
Moscow University of the Ministry of Internal Affairs named after V. Ya. Kikot, Moscow 17437,
Russian Federation
e-mail: sunobor@yandex.ru

© Springer Nature Switzerland AG 2019
A. G. Kravets (ed.), *Big Data-driven World: Legislation Issues and Control
Technologies*, Studies in Systems, Decision and Control 181,
https://doi.org/10.1007/978-3-030-01358-5_4

1 Introduction

Forecasting of vitally important for humanity processes occupies a special place in science. Forecasting of events with the highest probability was always appreciated especially in our age of dynamic processes. Scientific forecasting is made in two ways: (1) on the basis of extrapolation of their behavior in the past and present; (2) on the basis of statistical processing of experts' opinions on specific issues and areas of knowledge.

The chapter presents an innovative approach developed by Belyaev [1], the essence of which is the use of computer statistics and definition of the events of our interest based on the theory of probability and mathematical statistics. The theory of Yu. K. Belyaev is called "theory of resampling". This approach has already been applied in the interests of scientific research of forecasting aviation accidents, because the general set of catastrophes corresponding to certain parameters does not allow us to use the classical approaches of probability theory, due to its limitations. "The theory of resampling" is applicable to forecasting, with a certain probability, in terms of low statistics of the general aggregate of homogeneous events. Let us consider the option of applying the "theory of resampling".

2 Background

Suppose we have a limited sampling of a few tens of realizations $T_1, T_2, ..., T_n$ of some random value T with an unknown distribution function.

Then we define the sample average and dispersion by known formulas:

$$T_0^* = \frac{\sum_{i=1}^{n} T_i}{n}; \tag{1}$$

$$D_0^* = \frac{1}{n-1} \sum_{i=1}^{n} (T_0^* - T_i)^2. \tag{2}$$

In the future, the verification of compliance with the specified requirements for determining the probability is carried out by comparing the sample mean (less often selective variance), obtained on the basis of limited data of rare events, with the corresponding prescribed (normative) values. However, the point estimates (1) and (2) in this situation are random, and such a comparison is completely unjustified. What decision can be proposed to determine the probability of an event occurring against the background of a limited general aggregate of the process of interest to us?

Until recently, there was no answer to this question. But, with the advent and wide introduction of computer technology into the practice of carrying out statistical experiments, a new section of mathematical statistics—computer statistics—has begun to develop roughly and qualitatively. It became possible to conduct extensive statistical studies, make full use of modeling for the construction of artificial distributions, verification of certain theoretical assumptions and hypotheses. The new theory of carrying out statistical research focused on the use of modern computer technology has been implemented. Some stable artificial distributions are constructed using computers to asymptotically repeat the distributions of interest to researchers, which previously remained unknown. For example, using this approach it became possible to obtain a stable artificial distribution that repeats the distribution of random difference between mathematical expectation of a random operating time T on a single event of one process ($M[T]$) and the sample mean (1) obtained on the basis of limited data on rare events of our interest.

This result, in turn, allowed to estimate an unknown value $M[T]$ statistically strictly and, equally important, to determine the probability that the true (unknown) mathematical expectation $M[T]$ will be bigger or less than a predetermined value. This circumstance allowed us to take a fresh look at the very procedure for setting requirements for the possibility of predicting events.

First artificial distributions that were discussed above were obtained by Efron and Tibshirani [2]. However, their studies have not all been rigorously substantiated. Later, an elegant "theory of resampling" was proposed with strong evidence of obtaining artificial distributions and their convergence to unknown distributions of interest to researchers in a relatively simple way. We'll continue to rely on this theory, widely presented in [3–6].

3   Formal Problem Definition

So we have the following general scheme of data collection. Let $(\Omega, B(\Omega), P_0(\cdot))$ be a basic probability space, i.e. Ω will be a multitude of elementary events ω, $B(\Omega)$ will be σ—algebra of all events and $P_0(\cdot)$ will be the true distribution on $(\Omega, B(\Omega))$. Data collection can be understood as a sequence of experiments $\varepsilon_1, \varepsilon_2, \ldots$. The result of the i-th experiment can be represented as a point $x_i \in X_i$, where X_i is the multitude of values x_i. We consider x_i as a value of X_i for estimating random value which is measurable mapping Ω on X_i. The probability that $X_i \in B$ is specified as $P_{0,i}(B) = P_0(X_i(\omega) \in B)$, where we assume that $P_{0,i}(\cdot)$ belongs to some known family of distributions of the probabilities $B_i = (P_{\theta,i}, \theta \in \Theta)$ or, equivalently, the true parameter $\theta_0 \in \Theta$. Of course, we assume that any $\theta_0 \in \Theta$ uniquely determines $P_{0,i}(\cdot)$.

We shall use the following assumption: the true variables X_i, $i = 1, 2, \ldots$ are independent. Here we do not assume that X_i is identically distributed or that $X_i = X$, $i = l, 2, \ldots$

For each fixed n we consider $X_n = (x_1, x_2, \ldots, x_n)$ as a set of statistical data. Let $\hat{\theta}_n = S_n(x_1, x_2, \ldots, x_n)$ be a valid point estimate of θ_0, i.e. in some sense $\hat{\theta}_n \to \theta_0$ at $n \to \infty$.

If $\theta \subset R^r$, than $\theta = (\theta_1, \theta_2, \ldots, \theta_r)^T$ and θ_i are the actual values. In this case, it is a valid measurement of deviations $\underline{\hat{\theta}}_n$ from $\underline{\theta}_0$ at various $\underline{\hat{\theta}}_n - \underline{\theta}_0$, $n = 1, 2, \ldots$.

If $\theta \subset R^r$, than $\theta = (\theta_1, \theta_2, \ldots, \theta_r)^{\mathrm{T}}$ and θ_i are the actual values. In this case, it is a valid measurement of deviations $\hat{\theta}_n$ from $\underline{\theta}_0$ at various $\hat{\theta}_n - \underline{\theta}_0, n = 1, 2, \ldots$.

If we are interested in parameter $\underline{T}_0 = \underline{T}(\theta_0)$ where $\underline{T} : \Theta \to G, G = (\underline{T}(\theta_0) : \theta \in \Theta) \subset R^s$ than we can consider

$$L_{\theta,n} = L\left(\left(\hat{\underline{T}}_n - \underline{T}_n\right)\right), \tag{3}$$

where $\hat{\underline{T}}_n = \underline{T}_n(x_1, x_2, \ldots, x_n)$ is a point estimate $\underline{T}_0 = \underline{T}(\theta_0)$

Let $\bar{T}_n = \underline{T}_n\left(\hat{\theta}_n\right)$. If we consider $\underline{\theta}_n$ as the true parameter then $\underline{T}_n$ should be considered as the true value of the parameter of our interest.

In more general case, when we may consider $\underline{G}$ as a metric space with the metric of distance $d(\cdot, \cdot)$ the deviation equals to $d(\hat{T}_n, T_0)$, and we would like to know more about

$$L_{T(\theta_0),\,n} = L\left(\sqrt{n}d\left(\hat{T}_n, T_0\right)\right).$$

The question arises, whether it is possible to determine or at least to learn something more about the distribution (1), when $\hat{T}_n$ is unknown?

This question was answered in the affirmative in the above-mentioned works [2]. However, there is another more general and in some situations a more reasonable approach to reproducing the data [7]. Sometimes it seems natural to think due to the assumption of independence of $x_i (i = 1, 2, \ldots, n)$ that the order of conducting experiments has no significant effect in statistical conclusions, for example, to determine point estimate.

The same conclusion should take place when the experiments $\varepsilon_1, \varepsilon_2, \ldots, \varepsilon_n$ will be carried out in a different order, say $\varepsilon_{i_1}, \varepsilon_{i_2}, \ldots, \varepsilon_{i_n}$. Further, let the initial data be $(x_{i_1}, x_{i_2}, \ldots, x_{i_n})$. Here $\{i_1, \ldots, i_n\}$ is a permutation of $\{1, \ldots, n\}$.

Now $\varepsilon_1, \varepsilon_2, \ldots, \varepsilon_n$ is only a random sample from a more general set of possible experiments. Further, why not to multiply the data by sampling with the return from $\varepsilon_1, \varepsilon_2, \ldots, \varepsilon_n$? This would be equivalent to generating copies by statistics way by random sampling with replacement from the original statistics data $(x_{i_1}, x_{i_2}, \ldots, x_{i_n})$. We are interested in more knowledge about the distribution of the variance of point estimate $\bar{T}_n := T_n(x_1, x_2, \ldots, x_n)$ for the parameter of interest $\underline{T}_0 = \underline{T}(\theta_0)$.

Therefore we shall assume that $T_n(\cdot)$ will also be determined for resampling statistics data.

4   Resampling Method in Events Prediction

In [3] this approach is called resampling. Its exact definition is described by the following sequence of actions.

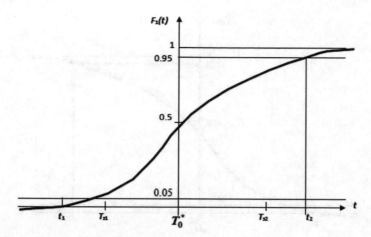

Fig. 1 The probability when $T_{n2} > T_{n1}$

1. We should find $\hat{T}_n = T(X_n)$ that is a consistent assessment of the parameter of interest.
2. We should take B of random samples with return of the original data of the same length of n: $X_n^{*1} = (x_1^{*1}, x_2^{*1}, \ldots, {}_n^{*1})$, $X_n^{*2} = = (x_1^{*2}, x_2^{*2}, \ldots, {}_n^{*2})$, ..., $X_1^{*B} = (x_1^{*B}, x_2^{*B}, \ldots, {}_n^{*B})$. We shall call X_n^{*b} a resampling copy of the originals.
3. If $\underline{T}_0 = \underline{T}(\theta_0)$ is the parameter we are interested in and $\hat{T}_n = \underline{T}(X_n)$ is a consistent evaluation, then each resampling copy of the original data is found as
4. $\hat{\underline{T}}_n^{*1} = T(X_n^{*1}), \hat{\underline{T}}_n^{*2} = T(X_n^{*2}), \ldots, \hat{\underline{T}}_n^{*B} = T(X_n^{*B})$.
5. We should find deviations of the estimated values $\hat{\underline{T}}_n^{*}$ from, i.e. $\left(\hat{\underline{T}}_n^{*1} - \hat{T}_n, \hat{\underline{T}}_n^{*2} - \hat{T}_n, \ldots, \hat{\underline{T}}_n^{*B} - \hat{T}_n, \right)$. After that let's estimate resampling ver-
6. sion for the conditional law of deviations from $\hat{T}_n$ specified via $(x_{i_1}, x_{i_2}, \ldots, x_{i_n})$:

$$L_{n,B}^{*} = L\left(\left(\hat{\underline{T}}_n^{*} - \hat{T}_n \right) | x_1, x_2, \ldots, x_n \right) = \frac{1}{B} \sum_{b=1}^{B} I\left(\left(\hat{\underline{T}}_n^{*} - \hat{T} \right) \in \bullet \right).$$

Under certain assumptions $L_{n,B}^{*}$ will repeat $L_{\underline{T}(\theta_0),n}$ for large values of n.

Based on the above resampling algorithm, a method was proposed for predicting the onset of a future event, determined from a limited number of statistical data of the general population of all homogeneous events [8, 9].

Let us dwell only on the interpretation of the final numerical results obtained in relation to the topic of this chapter, referring to Figs. 1 and 2.

Figure 1 shows the distribution function $F_1(t)$ of the random difference $T_0^* - M[T]$ obtained with the help of artificial resampling redistribution for a limited initial sampling. (Here the qualitative picture is presented, quantitative results see in [10, 11]).

From Fig. 1 it is seen that (here $\alpha = 0.90$)

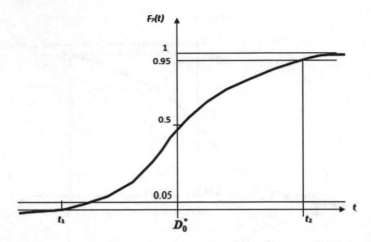

Fig. 2 The distribution function $F_2(t)$

$$P\{t_1 < M[T] < t_2\} = F_1(t_2) - F_1(t_1) = \alpha = 0.95 - 0.05 = 0.90.$$

Let's consider the probability with which the given requirements T_{31} and T_{32} ($T_{32} > T_{31}$, see Fig. 1) are fulfilled
Obviously,

$$P\{M[T] > T_{31}\} = F_1(t = \infty) - F(T_{31}) = 1 - F(T_{31}) = \alpha_1 \text{ and } P\{M[T] > T_{32}\}$$
$$= F_1(t = \infty) - F(T_{32}) = 1 - F(T_{32}) = \alpha_2.$$

It is clear that $\alpha_1 > \alpha_2$.

Therefore, the verification procedure for compliance of the true indicators of a foreseeable event should consist in:

– checking whether the requirement falls within a given significant interval α or to the left of it (the latter is the most favorable case);
– determination of the probability $P\{M[T] > T_3\}$ and comparing it with the value of α_3 (this is the main difference between the proposed approach and the specification of requirements for forecasting events);
– if $P\{M[T_3] > T_3\} > \alpha_3$ than the requirements for forecasting events are presented.

Obviously, the worst case (in the sense of satisfying the specified requirements) will be when the value T_3 is set at the right edge of the meaningful range (for example corresponding to $\alpha = 0.90$) or outside on the right.

Let us show, how the situation with the verification of the dispersion requirements is. Setting the requirements for variance in solving problems of predicting probabilistic events, when organizing and determining the accuracy of determining parameters, has a clearly expressed meaning. However, when solving security tasks, the requirements for the dispersion of proceeding of a homogeneous general population are not yet exhibited, and no applicable norms are provided for them.

Figure 2 shows the distribution function $F_2(t)$ of a random difference of sampling dispersion (obtained with a limited sampling of fixed volume) D_0^* and dispersion D $[T]$ of the population, i.e. the true dispersion. This function was also obtained using artificial resampling distribution (Only a qualitative picture is presented in Fig. 2).

If there is a need to determine with what probability the true value of the variance will fall into the range of interest (for example t_1, t_2), then this probability is

$$P\{t_1 < D[T] < t_2\} = F_2(t_2) - F_2(t_1) = \alpha.$$

According to Fig. 2 $F_2(t_1) = 0.05$; $F_2(t_2) = 0.95$, and for this case

$$P\{t_1 < D[T] < t_2\} = 0.90.$$

The values t_1, t_2 and α can be further compared with the corresponding preset values, if any of them will be reasonably exposed.

5 Conclusion

Procedures of testing for compliance of true probabilistic values that are set on the basis of a limited sampling for indicators of safety, reliability, survivability, and sustainability will be similar to those outlined above.

Creation of a single interdepartmental concept and a scientific and methodological base for information and analytical support for decision-making in the field of public security and law enforcement is an important strategic issue. The proposed approach, according to the authors' opinion, is promising and effective, despite its relative novelty. This way requires the creation of large information arrays, which will undoubtedly require the use of powerful computers and highly qualified specialists for their servicing, the allocation of additional material and financial resources.

The practical implementation of the proposed approach can be useful both for assessing possible threats and for predicting favorable and negative events. Wide application of the proposed methodology can lead to the minimization of predicted, probabilistic technogenic accidents and catastrophes.

References

1. Belyaev, Y.K.: Bootstrap, Resampling and Mallow Metric. Institute of Mathematical Statistic, Umea, Sweden. Lecture Notes # 1 (1995)
2. Efron, B., Tibshirani, R.J.: An Introduction to the Bootstrap. Chapman and Hall, London (1993)
3. Good, P.I.: Resampling Methods: A Practical Guide to Data Analysis. Birkhäuser, Basel (2006)

4. Good, P.I.: Permutation, Parametric, and Bootstrap Tests of Hypotheses. Series in Statistics. Springer, New York (2005)
5. Mitov, K.V., Omey, E.: Renewal Processes. Springer Briefs in Statistics. Springer International Publishing, London (2014)
6. Nakagawa, T.: Maintenance Theory of Reliability. Springer Series in Reliability Engineering. Springer, London (2005)
7. Wang, H., Pham, H.: Reliability and Optimal Maintenance. Springer Series in Reliability Engineering. Springer, London (2006)
8. Nakagawa, T.: Advanced Reliability Models and Maintenance Policies. Springer Series in Reliability Engineering. Springer, London (2008)
9. Birolini, A.: Reliability Engineering: Theory and Practice. Springer, Heidelberg (2014)
10. Betskov, A.V.: Development and Validation of Methods of Assessment of Safety Performance of Air Traffic in the Russian Federation Based on the Limited Initial Statistics. PhD thesis. Moscow State Technical University, Moscow (2002)
11. Bondarenko, Y.V., Azarnova, T.V., Kashirina, I.L., Goroshko, I.V.: Mathematical models and methods of assisting state subsidy distribution at the regional level. In: International Conference on Applied Mathematics, Computational Science and Mechanics: Current Problems, 18–20 Dec 2017. Voronezh. Russian Federation. https://doi.org/10.1088/1742-6596/973/1/012061

Theoretical and Applied Aspects of Orthogonal Coding in Computer Networks Technologies

Valery Makarov, Vladimir Gaponenko, Boris Toropov
and Alexander Kupriyanov

Abstract The issues of building secure communication channels do not lose their relevance. The development of new methods of noise-immune encoding, as well as the protection of transmitted information from unauthorized access by third parties, remains an important scientific and practical task. The chapter deals with the formation of a newly modified set of orthogonal signals, the mathematical models of which are the modified set of piecewise constant orthogonal functions of Rademacher and Walsh. The technique of forming a plurality of orthogonal signals for data transmission through communication channels and processing a complex composite multi-level sum signal is provided, the form of which displays the state of the parallel interface of the computer complex and its processing by the pseudo-correlation devices of the receiver. The received signal coding system in the communication channel allows one to solve both the problem of providing noise immunity and protecting the transmitted information from unauthorized perception and recognition, which potentially contributes to the overall effectiveness of the protection of transmitted information.

Keywords Orthogonal Walsh and Rademacher functions
Modification of orthogonal signals · Modulo two summaries
Pseudo correlation processing · Total orthogonal set
Resistance to unauthorized recognition · Compaction of signals in form

V. Makarov · V. Gaponenko · B. Toropov (✉)
Academy of Management of the Ministry of Internal Affairs of Russia,
Moscow 125993, Russian Federation
e-mail: torbor@mail.ru

V. Makarov
e-mail: ovorta@mail.ru

V. Gaponenko
e-mail: profgaponenko@gmail.com

A. Kupriyanov
Moscow Aviation University (National Research University), Moscow 125993, Russian
Federation
e-mail: aik@mai.ru

© Springer Nature Switzerland AG 2019   47
A. G. Kravets (ed.), *Big Data-driven World: Legislation Issues and Control
Technologies*, Studies in Systems, Decision and Control 181,
https://doi.org/10.1007/978-3-030-01358-5_5

1 Introduction

The use of orthogonal codes as one of the most effective methods for increasing the reliability of information processing was pointed out in the works of the Academician V.A. Kotelnikov. The main advantage of the methods of noise-immune coding is the detection and correction of errors arising from the effects of interference in code-words. This ability to detect and correct errors is achieved by introducing redundancy in the construction of code tables. Moreover, errors can be detected and corrected only within the limits restricted by the corrective ability of the code.

One of the methods of compaction and separation of channel signals and separate elements of orthogonal codes that allow not only to eliminate redundancy, but also to ensure high reliability of information processing is the use of orthogonal codes followed by their processing by receiving correlation devices. In their structure, such signals refer to complex composite signals whose base is much larger than one ($B = F_{max} * T >> 1$, where: F_{max} is the maximum frequency in the spectrum of the transmitted signal, T is the signal period) and which are a kind of noise-like signals.

When building tele-access systems to computational resources using orthogonal codes whose mathematical elements are sets of different orthogonal piecewise-constant or continuous orthogonal functions or polynomials, it is necessary to create a basis of primitive orthogonal functions or polynomials [1]. Although different telecommunication systems, based on orthogonal signals [2, 3], are now widely distributed all over the world, the problem of choosing a specific signals class and their effective generation is an actual scientific problem.

This is a reason of the increased popularity of orthogonal signals of this class in modern telecommunication systems (e.g. CDMA [4, 5], OFDM [6], UMTS, CDMA2000 and others [7–10]).

2 Choosing an Orthogonal Basis for Building a Data Transmission System

It is known that in the construction of multichannel data transmission systems with multiplexing and separation of channel signals in form, various orthogonal code sequences based on orthogonal functions or Legendre, Chebyshev, Laguerre, Hermite, Jacobi, Bessel, Gegenbauer, Rademacher, Haar, and Walsh polynomials are used. Of all the listed functions and polynomials, it is necessary to select only those that are most effective for generating channel signals or for individual elements of code combinations.

Thus, the Laguerre and Hermite polynomials are orthogonal in the interval $-\infty \ldots +\infty$ and $0 \ldots +\infty$ and the limitation of the message transmission period is associated with the violation of orthogonality, and, consequently, with the appearance of interference of the channel signals.

The functions of Gegenbauer and Jacobi satisfy the condition of finite limits of orthogonality, but their technical realization is associated with considerable difficulties. The Bessel functions of the first and second kinds are not completely orthogonal, and their application is also associated with the appearance of errors due to mutual influence.

The most acceptable functions as signal-forming ones are the orthogonal Chebyshev and Legendre polynomials and the orthogonal functions of Rademacher and Walsh. However, the technical and software realization of signals, whose mathematical models are orthogonal Chebyshev and Legendre polynomials, is also difficult because of the use of complex analog multipliers in transmitting and receiving devices.

The orthogonal Rademacher and Walsh functions are most suitable for constructing orthogonal signals and orthogonal codes. However, when choosing certain orthogonal functions and polynomials as mathematical models of orthogonal signals and orthogonal codes in the construction of tele-access systems to computational resources, it is necessary to be guided not only by the degree of complexity of their implementation, but also by the degree of exposure of such signals to different types of interference, and to unauthorized perception and recognition.

External disturbances are pulsed and fluctuation noise, "packet" noise, interference, concentrated in the spectrum or in time. The most stable to interference signals are signals in which the degree of compliance with interference is minimal. So, for the case of impulse noise prevailing in the tele-access channels, such an estimate can be made from the coefficients of approximation of the reaction of the communication line to the shock excitation from impulse noise. In this case, the reaction of the communication line to shock excitation can be expressed by a linear combination of mutually orthogonal functions if the latter form a complete basis.

For those functions for which the approximation coefficients will be minimal for the same number of terms of the sum of the approximating series, the correspondence between the functions describing information signals and interference will also be minimal. Consequently, the signals described by these orthogonal functions will be most resistant to the destructive interference effect.

$$\beta = \frac{\int_0^T U_{\bar{i}}(t) * U_{\bar{n}}(t)dt}{\int_o^T U_{\bar{n}}^2(t)dt}, \tag{1}$$

where: $U_n(t)$ is a system of orthogonal functions describing the interference; $U_c(t)$ is the system of orthogonal functions describing the useful signal.

An analysis of the existing methods of organizing a tele-access system to computational resources has shown that they mainly use time-based multiplexing and separation of channel signals, which loses in reliability to other methods of multichannel data transmission. In this regard, the most promising is the organization of tele-access with multiplexing and separation of channel signals in a form in which orthogonal codes based on orthogonal Walsh functions are used as channel signals, together with their optimal processing in receiving correlation devices.

In the system under consideration, a generalized complete system of orthogonal piecewise-constant Walsh functions is used to construct channel code-forming signals. This approach requires new qualitative changes in the construction of a general communication theory based on sinus-cosine functions and digital methods of information processing. In this case, the description of the signal processing methods does not occur in the frequency-time plane, but in the function-time plane.

Consequently, any sequence of orthogonal signals built on a complete system of orthogonal functions occupies a finite part of the functional-time plane.

By a complete orthonormal system of functions, we mean a system in which for any function $F_i(t)$ the limit of a quadratically integrable difference tends to zero. For such a system, the Bessel inequality $\sum_{k=1}^{\infty} F_m^2 \leq \|a\|^2$ in the passage to the limit becomes Parseval's equation. Such a representation allows us to evaluate in the physical sense the energy of non-sinusoidal oscillations as a sum of the energies of individual spectral components:

$$\int_0^T F^2(t)dt = \sum_{k=1}^{\infty} (a_k^2 + b_k^2) \tag{2}$$

This approach allows us to calculate the signal values at the outputs of individual correlation devices of multi-channel data transmission systems with channel separation in shape, the channel signals of which are constructed on the basis of non-sinusoidal complex composite orthogonal functions or polynomials.

The construction of the channel-forming equipment of computerized complexes, the protection of the converted and transmitted data using orthogonal signals based on sets of orthogonal functions or polynomials as elements of code combinations have not been widely used to date due to a relatively small degree of the study compared to the classical methods of processing data.

This chapter discusses the organization of a data transmission and protection system in computer technologies using methods of compaction and separation of elements of code combinations in a form whose mathematical models are the orthogonal Walsh functions related to the set of orthonormal piecewise constant orthogonal functions.

It should be noted that the system $\{f(j, x)\}$ of real and non-zero functions is said to be orthogonal at a finite interval $x_0 \leq x \leq x_1$, if the following conditions are satisfied:

$$\int_{x_0}^{x_1} f(j, x) \cdot f(k, x) = x_j \cdot \delta_{jk}, \delta_{jk} = \begin{cases} 1, & j = k \\ 0, & j \neq k \end{cases} \tag{3}$$

These orthogonality conditions are defined in the metric of the Hilbert space. In Euclidean space, the orthogonality condition is defined as:

$$f(j, x) \cdot f(k, x) = f(j, x) \cdot f(k, x) \cdot \cos \varphi = x_j \delta_{jk}$$

$$\delta_{jk} = \begin{cases} 1, \ \varphi = 0° \\ 0, \ \varphi = 90° \end{cases} \tag{4}$$

It follows from the expressions (3, 4) that the vectors describing the elements of signals or code combinations are orthogonal if their scalar product is equal to 1 in the case of their complete coincidence and 0 otherwise.

3 The Construction of a Signal System Based on the Set of Rademacher-Walsh Functions

In studies conducted by Walsh, the possibility of forming a complete system of orthogonal piecewise constant functions is shown on the basis of the fundamental orthogonal Rademacher functions, which are a subset of the complete piecewise constant set of orthogonal Walsh functions.

In such a system, the basic orthogonal Rademacher functions are functions of the form $M_r \Rightarrow \{Y_1, Y_2, Y_4, Y_8, Y_{16}, \ldots Y2^n\}$. A complete system of orthogonal Walsh functions is formed from the basic orthonormal system of piecewise constant Rademacher functions by their algebraic multiplication, for example, from Rademacher's orthogonal functions Y_1 и Y_2, the Walsh function Y_3 is defined as $Y_3 = Y_1 * Y_2$. Each subsequent derivative of the Walsh function is generated according to the algorithm:

$$Y_5 = Y_1 * Y_4; \ Y_6 = Y_2 * Y_4; \ Y_7 = Y_1 * Y_2 * Y_4;$$
$$Y_9 = Y_1 * Y_8; \ldots; Y_{15} = Y_1 * Y_2 * Y_4 * Y_8; \tag{5}$$

When determining the necessary and sufficient number of orthogonal Walsh functions for constructing individual elements of orthogonal codes in data protection and transmission systems, it is necessary to reveal the maximum lower limit of the number of the Rademacher base function sup K_j. After that, the basic functions of the selected subset are determined from which the complete system of the orthogonal Walsh subset is formed on the basis of multiplication. The sequence number of each such function is determined in accordance with the following formula:

$$m = sup \, K_j + \sum_{j=1}^{2^{n-1}} K_j \tag{6}$$

where: m is the ordinal number of the derivative of the Walsh function; K_j are the numbers of the basis generators of the Rademacher functions.

In turn, the ordinal number of the derivative of the Walsh function determines its structural composition, i.e. a subset of the generators of the Rademacher functions.

In general, the algorithm for forming signal-forming piecewise constant orthogonal Walsh functions is defined as

$$Y_m = Y_{sup\,Kj} * \prod_{j=1}^{2^n} Yk_j \{\forall Y_m, \, sup\,K_j + \sum_{j=1}^{2^n} K_j\} \qquad (7)$$

In accordance with the formula (3), a fragment of the set of orthogonal piecewise constant orthonormal Walsh functions for constructing code orthogonal code combinations consisting of 64 elements of orthogonal codes mapped in a piecewise constant orthogonal Walsh set is presented in Fig. 1.

It should be noted that this system differs from the classical system of orthogonal Walsh functions. The authors of the chapter have succeeded in obtaining a modified set of piecewise-constant Walsh functions in the mapping in which all complex constituents of individual Walsh functions are strictly phased, i.e. begin with one phase, which is of great importance, both for technical and software implementation of the complete set of code orthogonal signals in tele-access systems to computing resources.

This advantage is determined primarily by the fact that it does not require the introduction of special recognition units for the initial phase, as separate channel signals, as well as a complex composite total signal, whose form will determine the current state of the parallel interface of the computer complex.

As can be seen from Fig. 1, the complete system of orthogonal Walsh functions consists of even and odd functions, taking for the whole period the values "+1" or "−1". However, it is necessary to check the orthogonal functions obtained for the absence of the violation of their orthogonality. Verification of the Walsh functions obtained and constructed shows that whatever combination is taken, the orthogonality condition is completely preserved. The last assertion allows us to preserve the separability of the constructed set, and hence to realize the recognition and isolation of each element of the given set of piecewise-constant orthogonal functions. An example of mapping the formation of a Walsh subset from the Rademacher basis functions is presented in the form of a graph (Fig. 2), which shows the algorithm for the formation of derivative functions based on Rademacher basis functions (for the case Y_1; Y_2; Y_4; Y_8); derivative of the Walsh function is determined by the sum of the numbers of the Rademacher functions entering the node under consideration.

The transformed graph is characterized by the fact that each subsequent function is formed as a result of multiplying one of the basic and derived functions or two basic functions, i.e., in this case, each node of the graph has only two inputs. The obtained condition is very important for the technical realization of the generator of orthogonal oscillations because in this case, its construction is oriented at the use of two-input coincidence circuits in devices for multiplying orthogonal signals.

The last transformation is also important for the software implementation of the generator of orthogonal oscillations, whose mathematical models are the set of piecewise constant orthogonal Walsh functions. In this case, only two functions are mul-

Y1=-1-1-1-1-1-1-1-1-1-1-1-1-1-1-1-1-1-1-1-1-1-1-1-1-1-1-1-1-1-1-1-1-1-1-1+1+1+1+1+1+1+1+1+1+1+1+1+1+1+1+1+1+1+1+1+1+1+1+1+1+1+1+1+1+1+1+1

Y2=-1-1-1-1-1-1-1-1-1-1-1-1-1-1-1-1+1+1+1+1+1+1+1+1+1+1+1+1+1+1+1-1-1-1-1-1-1-1-1-1-1-1-1-1-1-1-1+1+1+1+1+1+1+1+1+1+1+1+1+1+1+1+1+1

Y3=+1+1+1+1+1+1+1+1+1+1+1+1+1+1+1+1-1-1-1-1-1-1-1-1-1-1-1-1-1-1-1-1-1-1-1-1-1-1-1-1-1-1-1-1-1-1-1-1+1+1+1+1+1+1+1+1+1+1+1+1+1+1+1+1

Y4=-1-1-1-1-1-1-1-1+1+1+1+1+1+1+1+1-1-1-1-1-1-1-1-1+1+1+1+1+1+1+1+1+1-1-1-1-1-1-1-1-1+1+1+1+1+1+1+1+1-1-1-1-1-1-1-1-1+1+1+1+1+1+1+1+1

....

Y8=-1-1-1-1+1+1+1+1-1-1-1-1+1+1+1+1-1-1-1-1+1+1+1+1-1-1-1-1+1+1+1+1-1-1-1-1+1+1+1+1-1-1-1-1+1+1+1+1-1-1-1-1+1+1+1+1-1-1-1-1+1+1+1+1

Y9=+1+1+1+1-1-1-1-1+1+1+1+1-1-1-1-1+1+1+1+1-1-1-1-1+1+1+1+1-1-1-1-1-1-1-1-1+1+1+1+1-1-1-1-1+1+1+1+1-1-1-1-1+1+1+1+1-1-1-1-1+1+1+1+1

....

Y16=-1-1+1+1-1-1+1+1-1-1+1+1-1-1+1+1-1-1+1+1-1-1+1+1-1-1+1+1-1-1+1+1-1-1+1+1-1-1+1+1-1-1+1+1-1-1+1+1-1-1+1+1-1-1+1+1-1-1+1+1-1-1+1+1

Y17=+1+1-1-1+1+1-1-1+1+1-1-1+1+1-1-1+1+1-1-1+1+1-1-1+1+1-1-1+1+1-1-1-1-1+1+1-1-1+1+1-1-1+1+1-1-1+1+1-1-1+1+1-1-1+1+1-1-1+1+1-1-1+1+1

.... •

Y32=-1+1-1+1-1+1-1+1-1+1-1+1-1+1-1+1-1+1-1+1-1+1-1+1-1+1-1+1-1+1-1+1-1+1-1+1-1+1-1+1-1+1-1+1-1+1-1+1-1+1-1+1-1+1-1+1-1+1-1+1-1+1-1+1

Y33=+1-1+1-1+1-1+1-1+1-1+1-1+1-1+1-1+1-1+1-1+1-1+1-1+1-1+1-1+1-1+1-1-1+1-1+1-1+1-1+1-1+1-1+1-1+1-1+1-1+1-1+1-1+1-1+1-1+1-1+1-1+1-1+1

....

Y62=-1+1+1-1+1-1-1+1+1-1-1+1-1+1+1-1+1-1-1+1-1+1+1-1-1+1+1-1+1-1-1+1-1+1+1-1+1-1-1+1+1-1-1+1-1+1+1-1+1-1-1+1-1+1+1-1-1+1+1-1+1-1-1+1

Y63=+1-1-1+1-1+1+1-1-1+1+1-1+1-1-1+1-1+1+1-1+1-1-1+1+1-1-1+1-1+1+1-1-1+1+1-1+1-1-1+1+1-1-1+1-1+1+1-1+1-1-1+1-1+1+1-1-1+1+1-1+1-1-1+1

Fig. 1 Fragment of 63 of orthogonal Walsh functions in binary form

tiplied by element multiplication, which drastically shortens the time of formation of the allowed orthogonal set of code elements.

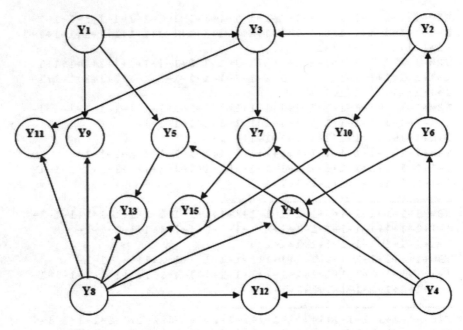

Fig. 2 An example of a transformed graph

In the sequential construction of a set of orthonormal Walsh functions, each successive function is constructed according to the following algorithm:

1. The first two orthogonal Rademacher functions Y_1 и Y_2 are constructed.
2. The function Y_3 is defined and constructed by the element-by-element multiplication.
3. Another Rademacher function Y_4 is introduced.
4. The functions Y_5, Y_6, Y_7 are constructed by the element-by-element multiplication of functions Y_1 и Y_4, Y_2 и Y_4, Y_3 и Y_4.

This process of complementing the Walsh set continues until a sufficient number of orthonormal functions are obtained to construct a plurality of code-forming orthogonal signals or elements of an orthonormal Walsh code.

4 Data Transmission System in the Orthogonal Basis

Temporarily fastening of the elements of the selected orthogonal set beyond the elements of the Windows code byte is conducted in accordance with the encoding table represented by $Y_k \rightarrow W_k$.

This fastening is dynamic and can change with each data transfer session. Due to the fact that orthogonal signals being models of orthogonal Walsh functions are parallel in time, it is possible to transmit not a sequence of Windows code bits but a complex composite multilevel sum signal consisting of a plurality of orthogonal Walsh signals and displaying the state of the parallel interface of the computer complex in a single time interval—T.

The latter makes it possible to form complex composite multilevel signal carrying information about the quantum of the transmitted data by means of the encoder. Such a semantic quantum may be composed of several symbols of the natural alphabet. In this example, such a semantic quantum is a word, part of a word, a phrase consisting of seven semantic elements (letters). About 64 orthogonal Walsh signals will be required to display any seven-digit combination of natural Cyrillic symbols, the digital alphabet and punctuation mark interpreted in Windows-codes.

Initially, any eight Walsh functions are assigned to the bits of the eight-element Windows code that represent the first symbol of the natural alphabet, then the eight orthogonal signals (taken from the remaining ones) are fixed after the bits of the second symbol of the encoded text, and so on. As a result of this conversion, all seven symbols of the encoded text are displayed on a set of orthogonal Walsh functions

$$\forall x_i \in X \to \{a_w\}|8 \to \{y_i \in Y\}|64 \tag{8}$$

Such a mapping is performed using the transformation operator, which establishes the correspondence rules between the Windows code elements and the elements of the set of orthogonal Walsh functions.

After the correspondence is established between the plurality of elements of the Windows code byte mapping the natural alphabet semantic symbol and the plurality of orthogonal Walsh signals from the correspondence schedule that is determined by the set and dynamically changing conversion table, the selected orthogonal signals are summed. As a result of this transformation, a complex composite multilevel signal is generated that displays the current state of the parallel interface of the computer complex. The block diagram of the orthogonal coding system is shown in Fig. 3.

The diagram in Fig. 4 shows a complex composite multilevel summary signal of the word "Krypton" consisting of 56 orthogonal Walsh signals ($Y_1 \ldots Y_{56}$).

3. The next step in converting parallel Windows code is to display the complex composite multi-level signal of the selected orthogonal set into binary code. To perform such a conversion, a dynamic conversion table is formed that establishes a correspondence between the individual levels of the total multilevel signal and the set of binary code words.

In the case of the selected allowed orthogonal Walsh set, the number of levels of the total signal representing the state of the parallel interface of the computer complex in the example under consideration will be 64 quanta. Moreover, the schedule for establishing correspondences between code combinations and quantized levels of a complex composite sum signal is set by the software method or is set by the user himself.

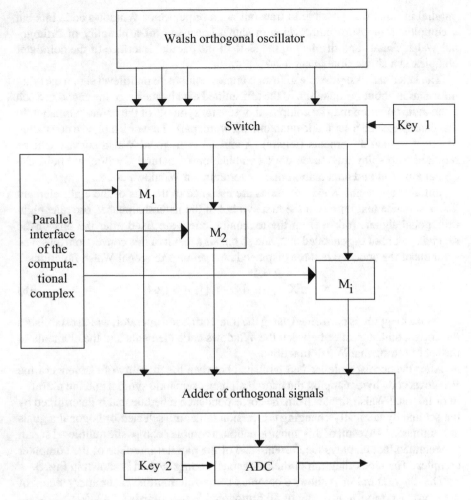

Fig. 3 Structural diagram of the orthogonal coding system

5 Conclusion

Considering the advantages of spread spectrum orthogonality, also it should be noted, that methods described are quite well matched to modern digital microchip circuitry (application-specific integrated circuit, very-large-scale integration, microprocessors). The main factor is the automatic acquiring of those benefits, following be the spread spectrum signals use, which cannot be seen directly within the classical reception framework but are numerous and very valuable in modern information systems. Achieved by complex multilevel sequence generation, rather than either time interval or bandwidth fragmentation, these codes provide interference immunity along with resistance to unauthorized access. Indeed, only the Key 1 and Key 2

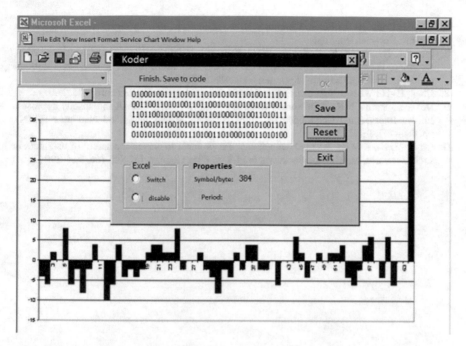

Fig. 4 Displaying the word "Krypton" with a complex composite multilevel signal

possession guarantees correct decoding of the complex signal generated and elimination of errors and distortions, which influenced the signal in the communication channel or computer system. This feature allows us to consider the system under study as a complex system of information transfer and protection.

This is a reason for the increased popularity of orthogonal signals of this class in modern telecommunication systems (e.g. CDMA, OFDM, UMTS, CDMA2000 and others).

References

1. Makarov, V.F.: The device for receiving television signals. Patent for Invention N2144741. Registered in the State Register of Inventions of the Russian Federation, 20 Jan 2000
2. Karanfıl, B., Tüysüz, B., Başaran, D.: Examination of the ambiguity function of multiple successive OFDM transmitters for passive detection applications. In: 25th Signal Processing and Communications Applications Conference (SIU), pp. 1–4 (2017)
3. Weidong, Kou: New orthogonal transform for digital signal processing. Electron. Lett. **21**(16), 666–667 (1985)
4. Fanucci, L., Luise, M., Giannetti, F., Rovini, M.: An Experimental Approach to CDMA and Interference Mitigation from System Architecture to Hardware Testing Through VLSI Design, p. 274. Springer, New York (2004)
5. Joseph, S., Samuel, M.: CDMA Radio With Repeaters, p. 424. Springer, New York (2007)

6. OFDM: Concepts for Future Communication Systems, p. 254. Springer, Heidelberg (2011). (Editor: Hermann Rohling)
7. Plass, S., Dammann, A., Kaiser, S., Fazel, K. (eds.): Multi-carrier systems & solutions 2009. In: Proceedings from the 7th International Workshop on Multi-Carrier Systems & Solutions. p. 404, Herrsching, Germany, Springer, Netherlands (2009)
8. Chen, H.-H.: What is Next Generation CDMA Technology. Wiley Telecom, 476 (2007)
9. Deka, S., Sarma, K.K.: Joint source channel coding with MC-CDMA in capacity approach. In: 4th International Conference on Signal Processing and Integrated Networks (SPIN), pp. 489–493, IEEE (2017)
10. Quyên, L.X., Kravets, A.G.: Development of a protocol to ensure the safety of user data in social networks, based on the Backes method. Commun. Comput. Inf. Sci. **466**, 393–399 (2014)

Part II
The Methodological Framework of Legislation Regulation in Big Data-driven World

Big Data in Investigating and Preventing Crimes

Elena Bulgakova, Vladimir Bulgakov, Igor Trushchenkov, Dmitry Vasilev
and Evgeny Kravets

Abstract The relevance of the research topic is caused by the discrepancy of the
modern capabilities of the Big Data technologies with the low rates of its implementation and use in the investigation and prevention of crimes. Today, Big Data
technologies are increasing their presence in solving problems in the system of public administration, economy, education, healthcare, ecology, construction, culture,
trade and other spheres of human activity. The limited use of the methods of analysis of Big Data in the investigation and prevention of crimes does not allow a
qualitative analysis of the available information of the crime, establish cause-effect
relations, create a social portrait of the criminal, which disadvantageously affects
the effectiveness of counter-crime activities. The chapter presents a comprehensive
analysis of the possibilities and directions of using the Big Data technologies in
solving typical problems of investigation and prevention of crimes. Some aspects of
using the capabilities of Big Data technologies in the investigation and prevention of
crime are given taking into account the comparative analysis of Russian and foreign
law enforcement experience. The main directions and conditions for the application of the Big Data technologies in the investigation and prevention of crimes are

E. Bulgakova (✉)
Kutafin Moscow State Law University (MSAL), 9 Sadovaya-Kudrinskaya str., Moscow 125993,
Russia
e-mail: oblaka7777777@gmail.com

V. Bulgakov · I. Trushchenkov
Moscow University of the Ministry of Internal Affairs of the Russian Federation
named after V.Y. Kikot, 12 Academician Volgin str., Moscow 117437, Russia
e-mail: vg.bulgakov@mail.ru

I. Trushchenkov
e-mail: hrustals@mail.ru

D. Vasilev · E. Kravets
Volgograd Academy of the Russian Ministry of Internal Affairs, 130
Istoricheskaya str., Volgograd 400089, Russia
e-mail: 89889599848@mail.ru

E. Kravets
e-mail: 80kravez@gmail.com

© Springer Nature Switzerland AG 2019   61
A. G. Kravets (ed.), *Big Data-driven World: Legislation Issues and Control
Technologies*, Studies in Systems, Decision and Control 181,
https://doi.org/10.1007/978-3-030-01358-5_6

determined and their perspectives are presented. The use of Big Data technologies in crime investigation and prevention will allow us to solve a wide range of crime prevention tasks at a higher level using professional expert systems and computer applications, as well as to improve the quality of crime investigation and reduce its terms.

Keywords Big data · Open data · Information technologies · The internet Information society · Multimodal evidence · System of public administration Computer network · Crime statistics · Electronic justice · Artificial intelligence

1 Background

The constant growth of the data volume [1–3], their updating the variety of types and methods of presentation, defines new tasks of collecting, processing, analyzing and managing such information. At present, the use of Big Data technologies is a universal tool for representing a new level of processing and analysis large amounts of information existing in modern society.

In Russia, Big Data technologies are already actively used in the public authorities [4], economics, education, health, sports, ecology, construction, culture, trade and other spheres of activity [5–15]. Unfortunately, there are still a few examples of the use of Big Data technologies in law, in particular for the prevention and investigation of crimes. For example, we can differ legal statistics [16]. In 2014 the report of the Secretary-General on improving the quality and accessibility of crime and criminal justice statistics indicates a shift to distributed data. In 2014–2015, the General Prosecutor's Office of the Russian Federation and the Open Government of the Russian Federation took a number of steps to increase the transparency of crime data. The methods of mathematical statistics and models are used in solving criminological problems. That makes it possible to obtain qualitative characteristics of criminality, to make forecasts on the development of the level of crime and manage it through decision-making. The data of criminal statistics are presented in "Open Data" format. The possibilities of modern methods of their analysis have expanded the circle of data consumers (economists, entrepreneurs, realtors, etc.). The ability to analyze distributed data has led to an increased demand to transform data into new knowledge [17–20].

Here are examples of the use of Big Data technologies in the organization of police activity in the United States of America. In particular, the Blue CRUSH (Crime Reduction Utilizing Statistical History) system, developed by IBM, is widely known and provides information for police about areas of the potential threat of crime. These data are based on the available crime statistics indicating the location (within a few blocks) and time for several hours of a specific day of the week) [21]. This kind of preventive forecasting led to a decrease in crime in Memphis by 31%, of which 15%–serious crimes [22]. Analytical solutions based on Big Data technologies are used in other US cities (New York, Seattle, Los Angeles, etc.) and the scale of their use increases every year [23].

2 Analysis of the Problems Connected with Using Big Data Technologies in the Crime Investigation

At the beginning of 1970 was created the first system in Russia of crime investigation, which was based on the use of the forensic characterization of a crime as an information model [24]. The structure of this information included the following elements:

- specification of the object of criminal assault;
- the environment for the commission of a crime;
- specification of the identity of the offender;
- specification of the victim's personality;
- typical ways of committing a crime;
- typical traces of the commission of a crime;
- the consequences of the crime.

On the basis of an accumulated data array were established the patterns of criminal activity, as a result of the investigation of similar crimes. Such an information model should reflect correlation and dependence together with significant links of its elements with the criminal's personality characteristics and the way he committed the crime.

The resulting report was to contain a set of characteristics of the offender and important circumstances for the criminal investigation.

Until recently, the imperfection of technologies and methods of obtaining and analyzing significant information did not allow the effective use of the forensic characteristic of crime.

The famous Russian scientist-criminalist Belkin [24] acknowledged that his own treatise–the criminalistic character of the investigation of certain types of crimes–it is a "phantom" due to the absence in that times technology that allows analyzing large amounts of data. Nowadays, the potential and opportunities of using the information model of crime–forensic characteristics–can be effectively implemented on the basis of Big Data technologies.

3 Directions for Using Big Data Technologies for Crime Investigation

Currently, there are opportunities for using Big Data technologies in the context of solving typical tasks that arise during the investigation and prevention of crimes.

Table 1 Typical tasks in crime prevention and investigation

Typical tasks	Big data technologies and methods of analysis
Analysis of legal information on the websites of state authorities	Statistical analysis Predictive analytics
Legal statistics	Statistical analysis
Judicial practice analysis	Statistical analysis Predictive analytics
Development of the information model of the crime	Statistical analysis Predictive analytics Simulation modeling
Analysis of electronic criminal cases on the identity of the perpetrator, at the place of commission, the manner of committing a crime and other grounds	Statistical analysis Predictive analytics Visualization of analytical data Data mixing and integrating
Detecting criminals using video surveillance databases of the automated information system "Safe city"	Visualization of analytical data Artificial neural networks Pattern recognition
The search for criminals on the basis of multi-modal biometric trace information systems [25]	Pattern recognition Artificial neural networks Data mixing and integrating
Creating a social portrait (profile) of the offender	Simulation modeling Visualization of analytical data Data mixing and integrating
Monitoring the criminal situation for the rational distribution of police forces and resources (resources)	Statistical analysis Predictive analytics
Prevention of corruption crimes using information technology	Crowdsourcing Predictive analytics
Assessment of public opinion on the effectiveness of law enforcement	Crowdsourcing Predictive analytics

Table 1 presents typical tasks in crime investigation, as well as Big Data technologies and methods used to solve them.

As a positive example of the implementation of Big Data technologies in the investigation and prevention of crimes in Russia, we will refer the comprehensive information system "Safe City" [26].

The modern city consists of many interconnected systems–transport, telecommunication, electricity, water supply, as well as many others. In addition, to the multistage and hierarchically complex management of all urban systems, it is still necessary to have adequate control over public law and order to ensure the security of every resident of the city. For this reason, Big Data technologies can be used with a greater degree of efficiency, allowing to receive and the process of differentiated information sets about all important events and transmit the received data for prompt response of competent services.

Big Data technologies can be used to solve video analytics problems. One of the elements of their use is modern security systems based on IP-video surveillance [27].

These systems provide automatic face detection, recognition, counting and object tracking for investigation. The software allows comparing faces with a database of stored video files in real time, for instance, for wanted criminals detection. There are already technologies for contactless identification in traffic, which is very important for the application in the field of law enforcement [28].

Most of the crimes are committed today on the Internet, which causes considerable difficulties in their investigation. The high level of latency of such crimes as drug trafficking, pedophilia, corruption, fraud, and others makes it difficult to identify them, carry out an objective investigation and obtain evidence. The use of Big Data technologies with the use of data mining and predictive analytics makes it possible to identify the motives for committing such crimes, establish the cause-effect relationships of various elements of the commission of a crime, and establish characteristics of the perpetrators.

The Criminal Code of the Russian Federation in 2017 was amended to establish criminal liability for the incitement to commit suicide using the Internet (paragraph 3 of Article 110.1 of the Criminal Code "Declination to commit suicide." [29] In the authors' opinion, the use of Big Data technologies will help identify those who commit such crimes, using methods of social networks monitoring, text messages, photos, video recordings, thematic avatars, specific symbols and other data analyzing.

Monitoring the crime situation with the use of Big Data technologies will allow identify predictable patterns of criminal activity and develop measures to prevent crimes based on long-term statistics and taking into account various situational factors.

With the use of Big Data technologies, it is possible to predict the behavior of a criminal by automatic analysis of all available information and drawing up his psychological portrait. The ability to predict the behavior of the perpetrator provides law enforcement authorities with a significant superiority in criminal activity.

For instance, this makes it possible to rationally use the forces and means on the basis of the knowledge of the further actions of the criminal, ensure the rapid detention of the offender, and not expose the victims to the risks.

At present, it is necessary to increase the share of public-private partnerships for solving the tasks facing law enforcement agencies. It is necessary to adapt Big Data technologies used in business and other spheres of human activity to solve problems of recognition and prevention of crimes [30, 31].

4 Requirements for Data Used in the Investigation and Prevention of Crime

The basic requirements for data for subsequent processing and use in investigation and prevention of crimes are:

- completeness;
- relevance;
- cumulative;
- precedence;
- timeliness;
- availability;
- saving;
- suitability for using;
- non-discrimination to access;
- lack of proprietary formats;
- licensing and others.

In our opinion, the list of basic requirements for data used in solving the tasks of investigating, include:

- confidentiality,
- relevance,
- reliability,
- authenticity,
- legality,
- verifiability and others.

In the criminal justice, certain requirements are imposed on electronic documents that are attached as evidence in a criminal case, such as relevance, admissibility, reliability, and sufficiency. In this regard, additional methods are needed to analyze the sources of information to comply with the requirements of the Russian Federation legislation of the collection, analysis, and use of personal data.

5   The Main Problems with Using Big Data Technologies in the Investigation

In our view, the main problems in the use of Big Data technologies in solving task of investigation are: the complexity of implementing the basic stages of the development of the information society; the protracted period of data converting to the Open data format; unresolved issues of legal and organizational support; editing and storing a large volume of constantly updated information, etc.

Currently, a significant problem is the disparity of data storage systems in various departments creates a barrier to access to this information. The desire to open as much data as possible to make quality and timely decisions is understandable. But it should not be forgotten about the possible negative consequences of disclosing personal information. We believe that one of the most pressing problems of using Big Data technologies in jurisprudence is the insufficiently developed mechanism of data classification and the legal regime of access to them. In particular, problems of observing the Federal Law of July 27, 2006, No. 152-FZ "About Personal Data" arise in connection with the application of Big Data technologies [32]. Thus, the use of Big Data technologies contradicts the principle of limiting personal data processing with predefined goals specified in Art. 5 of this Law. It is necessary to demand informed, concrete and conscious consent of the subject of personal data; depersonalization of personal data (for example, in the GAS Justice system), which does not secure the person from the possibility of opening them as a result of correlations with other databases.

Most of the data created between 2012 and 2020 will not be generated by people, but by different kinds of devices during their interaction with each other through data networks (for instance, smartphones, radio frequency identification devices RFID, satellite navigation systems such as GLONASS or GPS, etc.) [33]. At the same time, it is necessary to carefully analyze the current legislation, take into account the practice of foreign countries, to prevent precedents related to the violation of citizens' rights. The basis of this is to develop a data classifier that will be either publicly available or converted into the «open data» (in the understanding the Federal Law of July 27, 2006 No.149-FZ "About information, information technology and information protection"), or referred to the data of limited access (with the possibility of the legal data access and processing) [34]. The symbiosis of data of general access and limited access is important in solving specific tasks, for example, crimes investigation.

6 Conclusion

As a result, we will determine the list of priority infrastructure tasks that should be implemented to create conditions for the application of Big Data technologies in the investigation and prevention of crimes:

– decision-making system for police and others law enforcement;
– analytical data processing and regularities detection systems;
– systems of organization and management of all evidence-connected data;
– required IT infrastructure.

In summary, the use of Big Data technologies in the process of investigation is promising. If the conditions discussed in the chapter are met, the application of Big Data technologies will make it possible to more effectively solve the tasks assigned to law enforcement agencies for the investigation and prevention of crimes.

References

1. Gantz, J., Reinsel, D.: The Digital Universe in 2020: Shadows, and Biggest Growth in the Far East (2012). IDC I V I E W. http://www.emc.com/collateral/analyst-reports/idc-the-digitaluni verse-in-2020.pdf. Accessed 1 May 2018
2. Mohanty, H., Bhuyan, P., Chenthati, D.: Big Data, 184 p. Springer, India (2015). https://doi.or g/10.1007/978-81-322-2494-5

3. King, S.: Big Data. Potential und Barrieren der Nutzung im Unternehmenskontext, 182 p. Springer Fachmedien, Wiesbaden (2014). https://doi.org/10.1007/978-3-658-06586-7
4. Bulgakova, E.V., Bulgakov, V.G., Akimov, V.S.: The use of «Big Data» in the system of public administration: conditions, opportunities, prospects. Yuridicheskaya nauka I praktika: Vestnik Nizhegorodskoj akademii MVD Rossii–Legal Science and Practice: Bulletin of Nizhny Novgorod Academy of the MIA of Russia, 3(31), 10–14 (2015)
5. Open Data Russia. http://data.gov.ru//. Accessed 24 May 2016
6. Chen, M., Mao, S., Zhang, Y., Leung, V.C.: Big Data Related Technologies, Challenges and Future Prospects, 89 p. Springer Briefs in Computer Science, Springer International Publishing, Switzerland (2014) https://doi.org/10.1007/978-3-319-06245-7_1
7. Fasel, D., Meier, A.: Big Data. Grundlagen, Systeme und Nutzungspotenziale, 380 p. Springer Vieweg (2016). https://doi.org/10.1007/978-3-658-11589-0
8. Olteanu, D., Gottlob, G., Schallhart, C.: Big Data: Proceedings of 29th British National Conference on Databases, BNCOD 2013, 8–10 July 2013, 303 p. Springer-Verlag Berlin Heidelberg , Oxford, UK (2013). https://doi.org/10.1007/978-3-642-39467-6
9. König, C., Schröder, J., Wiegand, E.: Big Data. Chancen, Risiken, Entwicklungstendenzen. VS Verlag für Sozialwissenschaften, 178 p (2018). https://doi.org/10.1007/978-3-658-20083-1
10. Azarmi, B.: Scalable Big Data Architecture: A Practitioners Guide to Choosing Relevant Big Data Architecture, 141 p. Apress, New York (2016). https://doi.org/10.1007/978-1-4842-1326-1
11. Skourletopoulos, G., Mastorakis, G., Mavromoustakis, C., Dobre, C., Pallis, E.: Mobile Big Data: A Roadmap from Models to Technologies, 347 p. Springer International Publishing, Switzerland (2018). https://doi.org/10.1007/978-3-319-67925-9
12. Angelov, P., Manolopoulos, Y., Iliadis, L., Roy, A., Vellasco, M.: Advances in big data. In: Proceedings of the 2nd INNS Conference on Big Data, 23–25 Oct 2016, 348 p. Springer International Publishing, Thessaloniki, Greece (2017). https://doi.org/10.1007/978-3-319-47898-2
13. Pyne, S., Rao, B.L.S.P., Rao, S.B.: Big Data Analytics: Methods and Applications, 276 p. Springer, India (2016). https://doi.org/10.1007/978-81-322-3628-3
14. Tan, Y., Shi, Y.: Data mining and big data. In: Proceedings of First International Conference, DMBD 2016, Bali, Indonesia, 25–30 June 2016, 569 p. Springer International Publishing, Switzerland (2016). https://doi.org/10.1007/978-3-319-40973-3
15. Kolany-Raiser, B., Heil, R., Orwat, C., Hoeren, T.: Big Data und Gesellschaft. Eine multidisziplinäre Annäherung. VS Verlag für Sozialwissenschaften, 430 p (2018). https://doi.org/10.1007/978-3-658-21665-8
16. The Portal of Legal Statistics of the Prosecutor General of the Russian Federation. http://crimestat.ru/. Accessed 4 May 2016
17. Klous, S., Wielaard, N.: We are Big Data: The future of the Information Society, 199 p. Atlantis Press, France (2016). https://doi.org/10.2991/978-94-6239-183-3
18. Quinto, B.: Next-Generation Big Data: A Practical Guide to Apache Kudu, Impala, and Spark, 557 p. Apress, New York (2018). https://doi.org/10.1007/978-1-4842-3147-0
19. Zomaya, A.Y., Sakr, S.: Handbook of Big Data Technologies, 895 p. Springer International Publishing, Switzerland (2017). https://doi.org/10.1007/978-3-319-49340-4
20. Srinivasan, S.: Guide to Big Data Applications, 565 p. Springer International Publishing, Switzerland (2018). https://doi.org/10.1007/978-3-319-53817-4
21. Predictive Crime Fighting. http://www-03.ibm.com/ibm/history/ibm100/us/en/icons/crimefighting/transform/. Accessed 4 May 2016
22. Thompson, T.: Crime software may help police predict violent offences. In: The Guardian, July 25, 2010. http://www.theguardian.com/uk/2010/jul/25/police-software-crime-prediction. Accessed 4 May 2016
23. Joh, E.: Policing by numbers: big data and the fourth amendment. Wash. Law Rev. 89, 35–68 (2014)
24. Belkin, R.S.: Forensic Science: Problems of Today. Topical Issues of Russian criminology, 240 p. Publishing House NORMA, Moscow (2001)

25. Jain, A.K., Ross, A.A., Flynn, P.: Handbook of Biometrics, 556 p. Springer LLC, New York (2008)
26. Bulgakova, E.V., Akimov, V.S., Bulgakov, V.G.: Mechanisms of the electronic state for counteracting corruption. Leg. Inf. **1**, 23–27 (2014)
27. Intelligent Solutions Supporting IP Video. Axis Communications. https://www.axis.com/en-us/solutions-by-application. Accessed 6 July 2018
28. Motion Identification: Biometrics Technology Provides Safety and Freedom to Move. Axis Communications. https://www.axis.com/en-us/solutions-by-application/facial-recognition. Accessed 7 July 2018
29. The Criminal Code of the Russian Federation of June 7, 1996, No. 63-FZ (as amended on June 27, 2018). https://fzrf.su/kodeks/uk/. Accessed 7 July 2018
30. Quick, D., Choo, K.-K.R.: Big Digital Forensic Data. Volume 1: Data Reduction Framework and Selective Imaging, 96 p. Springer, Singapore (2018). https://doi.org/10.1007/978-981-10-7763-0
31. Quick, D., Choo, K.-K.R.: Choo, K-K. R.: Big Digital Forensic Data. Volume 2: Quick Analysis for Evidence and Intelligence, 86 p. Springer, Singapore (2018). https://doi.org/10.1007/978-981-13-0263-3
32. Federal Law No. 152-FZ of July 27, 2006 (as amended on 31 Dec 2017) "About Personal Data"
33. White, T.: Hadoop: The Definitive Guide, 3rd ed, p. 2. O'Reilly Media, Inc., CA, USA (2012)
34. Federal Law No. 149-FZ of July 27, 2006 (as amended on 29 June 2018) "About Information, Information Technologies and Information Protection"

Data Analysis of the Socio-economic Factors' Influence on the State of Crime

Igor Goroshko, Boris Toropov, Igor Gurlev and Fyodor Vasiliev

Abstract In the light of the emergence and rapid development of the digital economy, the development of mathematical models and methods for analyzing and forecasting the trends in the development of all possible spheres of public life comes to the forefront. In this connection, the analysis of the impact of socio-economic factors on the criminal situation in the regions of the Russian Federation is extremely topical. The research is based on the methods of variance analysis, correlation analysis, and regression analysis. The application of this mathematical apparatus to the data sets characterizing the socio-economic and criminal situation in the regions will promote a qualitatively new level of comprehension of how the aggregate of socio-economic indicators determines the criminality and its characteristics. Obviously, the models obtained by the authors do not pretend to exhaustively describe the numerical characteristics of criminality but are aimed at forming a methodological base for analyzing the criminal situation with the consideration of the social and economic situation.

Keywords Data analysis · Crime state · Mathematical models
Regression analysis · Correlation · Cluster analysis

I. Goroshko · B. Toropov (✉) · I. Gurlev · F. Vasiliev
Academy of Management of the Ministry of Internal Affairs of Russia,
Moscow 125993, Russian Federation
e-mail: torbor@mail.ru

I. Goroshko
e-mail: garrygo@mail.ru

I. Gurlev
e-mail: gurleff@mail.ru

F. Vasiliev
e-mail: vasilev17@mail.ru

© Springer Nature Switzerland AG 2019   71
A. G. Kravets (ed.), *Big Data-driven World: Legislation Issues and Control
Technologies*, Studies in Systems, Decision and Control 181,
https://doi.org/10.1007/978-3-030-01358-5_7

1   Introduction

Crime is a complex social phenomenon. Scientific definition of crime itself is an actual problem [1], lying in the fields of law, sociology and economy. What is obvious, crime rate in society depends on a large number of factors that determine it [2–4]. In this regard, a qualified specialist needs not only to have clear ideas about the main directions of the development of crime but also be able to take into account the complex interrelated variety of factors that have a significant impact on it. Such investigations cannot be carried out without knowledge of the foundations of probability theory and mathematical statistics, i.e. disciplines, allowing the researcher to understand a huge amount of stochastic information and to choose the only model that best reflects the phenomenon being studied.

It should be emphasized that modern methods of statistical data analysis are being actively used in various fields of human activity, including in the management of enterprises and organizations, allowing us to make fairly accurate forecasts about the state of the production environment, commodity markets, and price dynamics.

It seems that these methods can be useful for the activities of the bodies of internal affairs, especially as information technology has made them more accessible. The most laborious work on the calculation of various statistics and model parameters is mainly performed by a computer, and the analyst's task is setting the goals, justifying approaches to their achievement, and interpreting the results.

2   Methodology and Initial Data

The methodological basis for building crime models is the apparatus of cor-relation and regression analysis of data. Although these methods are widely presented in scientific and educational literature, such as [5–7], their use in each specific case is a new scientific task. The elements of cluster analysis [8] used to select and classify the observations that formed the basis of the models obtained. The clustering methods are now widely used for spatial crime data analysis [9, 10] and law enforcement management [11].

The basic information collection of this study is composed of the official da-ta on socio-economic development and the state of crime in the regions of the Russian Federation (with the exception of the Chechen Republic, the Republic of Crimea and Sevastopol) [12]. In addition, statistical information on the indicators of the Khanty-Mansiysk Autonomous County—Ugra, and the Nenets Autonomous County were included in the information on the Tyumen and Arkhangelsk Regions, respectively. The total number of regions considered in the guidelines is eighty.

3 Selection of Socio-economic Factors Affecting the State of Crime

The list of socio-economic factors affecting the state of the criminogenic situation in the regions of the Russian Federation and crime rates was formed as a result of a preliminary criminological analysis. It is correlated with approaches used by domestic as well as foreign economists, sociologists, and criminologists [2, 3].

Socio-economic factors were divided into two main groups.

The first group is composed of economic factors. The group of economic factors includes:

- average annual number of employees in the economy;
- per capita monetary income (per month), rubles;
- consumer spending on average per capita (per month), rubles.
- average monthly nominal accrued wages of employees of organizations, rubles;
- the gross regional product, million rubles;
- fixed assets in the economy (at full cost, at the end of the year), million rubles;
- the volume of shipped goods of own production, jobs performed and services rendered by one's own means in various types of economic activity, million rubles (mining);
- the volume of shipped goods of own production, jobs performed and services rendered by one's own means in various types of economic activity, million rubles (manufacturing industries);
- agricultural output—total, million rubles;
- commissioning of the total area of residential buildings, thousand square meters, m;
- turnover of retail trade, million rubles;
- investments in fixed assets, million rubles.

The second group comprises the factors that characterize the state of human development, which is determined, in particular, by the consumption of "harmful goods".

This group consists of the following indicators:

- the number of patients under observation with the first diagnosis of "alcoholism" and "alcoholic psychosis", for the first time in their lives, people;
- the number of patients taken under observation for the first time in life with the established diagnosis of "drug addiction", people.

In addition to the above, factors that to some extent characterize the processes of urbanization in the regions, namely, the number of the urban population (people) and demographic processes (the number of young people in the general structure of the population, people) have been included in the list of socio-economic factors.

The inclusion of the urban population in the study is explained by the need to test the hypothesis formulated by some scientists about the significant influence of this factor on victimization rates.

As for youth, this factor is traditionally used in criminology in studying contemporary trends in the development of crime.

Youth is a special socio-demographic group, special in its age, psychological characteristics, social status, etc. Age ranges for young people in Russia are identified as 15–29 years.

It is the representatives of this group that most clearly demonstrate their correspondence with the notion of "minors", which is reflected in the statistics of offenses (juveniles are teenagers who at the time of the crime commission were fourteen, but not eighteen).

The inclusion of various *indicators of crime* in the list of the investigation was made by several criteria.

By the criterion of the lowest latency, the following indicators were used in this research:

- the number of crimes registered under the Article 105 of the Criminal Code of the Russian Federation;
- the number of crimes registered under the Article 111 of the Criminal Code of the Russian Federation;
- the number of crimes registered under the Article 161 of the Criminal Code of the Russian Federation;
- the number of crimes registered under the Article 162 of the Criminal Code of the Russian Federation.

According to the criterion of the greatest prevalence, the crimes envisaged in the Article 158 of the Criminal Code were chosen.

In addition, such socially important indicators of crime as the number of juveniles who have committed crimes and the number of recorded crimes committed by the repeated criminals were included in the investigation.

In conclusion, we'd like to note that, like the list of socio-economic indicators, the list of crime indicators is largely consistent with the analyzed characteristics, which were most often brought to the attention of scientists-criminologists.

4   Clustering of Regions

The implementation of the clusterization procedure is explained by the large variety of Russian regions (differing in population size, socio-economic indicators, crime rates, etc.) and the need, in this connection, to verify the possibility of their integration into a single array for the subsequent analysis.

The task of clustering consists in splitting the set of investigated objects (in our case, these are regions) into groups that are homogeneous in terms of quantitative similarity with respect to the selected one or several characteristics (in other words, from the initial set of objects it is necessary to form several groups which are homogeneous by some criteria).

Before implementing such a breakdown, the analyst, based on his own experience and understanding of the essence of the problem, must admit that splitting into groups will not introduce significant distortions and will not violate the integrity of the statistical aggregate of objects.

Moreover, he should make sure that the chosen units of measurement of the compared characteristics (or the measurement scales) are comparable.

Observance of these conditions is a necessary precondition for the qualitative cluster analysis.

The rule of assigning an object to a particular group is based on the concept of "distance", measured either between each pair of objects x_i and x_j, or between the object under consideration and the selected center of the group. If the measured distances are close to or do not exceed a predetermined threshold, the objects are recognized as homogeneous and belonging to the same group.

In practice, the so-called Euclidean distance is most commonly used as a measure of closeness between objects, calculated for each pair of objects x_i and x_j by the formula:

$$\rho_\varepsilon(x_i, x_j) = \sqrt{\sum_{k=1}^{n} (x_{ik}, x_{jk})^2},$$ (1)

where k is the number of attributes chosen for clustering ($k = 1, 2, \ldots, n$).

In the proposed methodology, the Euclidean distance is calculated between the neighboring regions located in descending order of the size of the population residing in their territory (Table 1).

Signs are represented by the pairs of "population" and one of the indicators of crime (included in the list and formed at the first stage) acting in turn. Table 2 shows a fragment of the results of calculating the Euclidean distance for the signs "population size" and "number of crimes registered under the Article 105 of the Criminal Code of the Russian Federation", performed using the Excel table processor.

Table 1 Fragment of ranking of regions in descending order of population size

No.	Region	Population	The number of crimes registered under the Article 105
18	Moscow	12,197,596	373
10	Moscow region	7,231,068	574
28	St. Petersburg	5,191,690	256
56	Sverdlovsk region	4,327,472	457
40	Rostov region	4,242,080	273
44	Republic of Tatarstan	3,855,037	289
59	Chelyabinsk region	3,497,274	357
…	….	….	…

Table 2 Fragment of calculation of Euclidean distance for the signs "population size" and "number of crimes registered under the Article 105 of the criminal code of the russian federation"

No.	Region	Population	The number of crimes registered under the article 105	Euclidean distance between the objects
18	Moscow	12,197,596	373	4966528.004
10	Moscow region	7,231,068	574	1777739.021
28	St. Petersburg	5,191,690	256	864218.0234
56	Sverdlovsk region	4,327,472	457	85392.19824
40	Rostov region	4,242,080	273	170093.0389
44	Republic of Tatarstan	3,855,037	289	357763.0065
59	Chelyabinsk region	3,497,274	357	227071.0301
...	...	...	...	...

The statistical bases of cluster analysis, as well as the subsequent construction of spatial models, are the data for 2014, which are least susceptible to statistical outliers and gross errors than in subsequent years since this period was characterized by a more stable external economic environment.

After calculating the Euclidean distance, it is necessary to construct a frequency distribution of the results obtained in order to calculate the mean μ and standard deviation σ and determine, with their help, the homogeneity boundary of the objects by the criterion

$$|\rho_{\varepsilon(xi,xj)}| \leq \mu + \sigma \qquad (2)$$

The graph of the frequency distribution of the Euclidean distance for the Russian regions is shown in Fig. 1.

Figure 1 shows some shift of the frequency distribution to the left edge, so to find the mean and standard deviation values one should use the lognormal distribution formulas.

In the course of the calculations, the following values are obtained:

- mean $\mu = 1472396.642$;
- standard deviation $\sigma = 2951911.756$.

Further, by verifying the fulfillment of the condition (2) for each pair of neighboring regions, it is necessary to identify the homogeneity of the investigated collection and the possibility of uniting regions into a single group for further analysis.

In our calculation, (according to the criteria chosen for clustering, "population size" and "number of crimes registered under the Article 105 of the Criminal Code

Fig. 1 The graph of the
frequency distribution of the
Euclidean distance

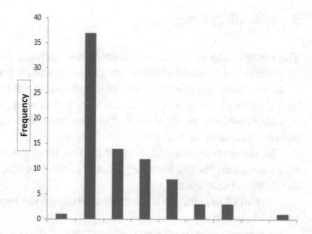

of the Russian Federation") all the surveyed regions, with the exception of Moscow, satisfy condition (2).

The Euclidean distance between the regions closest in population—Moscow and the Moscow region—exceeds the threshold value $(4966528.004 > \mu + \sigma)$, according to which objects are recognized as homogeneous and belonging to the same group.

Similar results were obtained for the remaining pairs of characteristics:

- "population size" and "number of crimes registered under the Article 111 of the Criminal Code of the Russian Federation";
- "population size" and "number of crimes registered under the Article 1158 of the Criminal Code of the Russian Federation";
- "population size" and "number of crimes registered under the Article 161 of the Criminal Code of the Russian Federation";
- "population size" and "number of crimes registered under the Article 162 of the Criminal Code of the Russian Federation";
- "population size" and "number of identified juveniles who committed crimes";
- "population" and "the number of recorded crimes committed by persons who previously committed crimes."

Thus, based on the preliminary analysis, it is concluded that the region of Moscow differs significantly from other regions of the Russian Federation according to the selected features and the need for its independent study. Further research is needed for seventy-nine regions. Further research is needed for seventy-nine regions. Clustering procedure defined, as tough, as other more complex clustering methods may be extrapolated from state to the regional and city level, as shown for example in [9–11].

5   Modeling Procedure

The construction of mathematical models describing the quantitative relationships between socio-economic factors and crime indicators is carried out in stages [5, 7].

At the first stage, the set of variables included in the model is determined, and the hypothesis and purpose of the research are formulated.

In our research, we'll consider all sixteen socio-economic factors and seven crime indicators defined by us above.

The research hypothesis is based on the assumption that the above-mentioned socio-economic factors have a significant effect on crime rates in selected regions of the Russian Federation.

The aim of the study is to confirm or disprove the hypothesis with a 95% significance level.

Further—at the a priori stage—the essence of the studied set of objects is analyzed and arrays of known prior to the beginning of the research (a priori) information are formed.

These information arrays were formed on the basis of data on the state of crime and the results of social and economic development in selected regions of the Russian Federation in 2014, obtained from the materials of state statistical reporting, formed by the Ministry of Economic Development of the Russian Federation, the Federal State Statistics Service and the General Information-Analytical Centre of the Ministry of Internal Affairs of Russia. The initial information series comprises 1840 items.

The third stage—*the stage of parametrization*—is devoted to choosing the general form of the model and determining the composition of its variables and parameters.

The experience of the authors in the field of application of mathematical methods in the analysis of data made it possible to dwell on the use of the mathematical *model of linear multiple regression* as a model describing the influence of socioeconomic factors on crime rates:

$$Y = b_0 + b_1 X_1 + b_2 X_2 + \cdots + b_n X_n + \varepsilon, \tag{3}$$

where Y—*dependent* (*explained*) variable, one of the indicators of crime; $X_1, X_2, \ldots,$ X_n—*independent* (*explanatory*) variables (socio-economic factors that affect Y); $b_0,$ $b_1, \ldots b_n$ are the parameters of the model, n is the number of independent variables, ε is the *random component* (*or remainder*) whose presence in the model is the result of ignorance or lack of data on all explanatory variables that somehow affect Y.

Note that the variables Y and X are column vectors of size m, where m is the number of observations.

An important task that needs to be solved at the parameterization stage is the task of selecting factors for inclusion in the model. It is recommended that the factors included in the mathematical model of multiple regressions meet the following requirements:

– they must be quantitatively measurable;

– factors should not be closely related to each other, in other words, they should not be multi-collinear;
– As practice shows, the inclusion of multi-collinear factors in the model leads to a number of negative consequences. In particular:
– it may be difficult to interpret the parameters of multiple regression as characteristics of the impact of factors on crime rates in a "pure form";
– parameter estimates become unreliable in a statistical sense because they have large standard errors and are subject to variations, depending on the number of observations.

To calculate the strength of the connection between the socio-economic factors xi and xj, entering into each of the groups defined in Sect. 3, the pair correlation coefficients rxi, xj are used.

Using the "Correlation" tool in the Excel "Data Analysis" module, we can construct a correlation matrix whose cells contain numerical values of the correlation coefficients.

We'll use the following indications:

– average annual number of employees in the economy—X1;
– per capita monetary income (per month), rubles—X2;
– consumer spending on average per capita (per month), rubles—X3;
– average monthly nominal accrued wages of employees of organizations, rubles—X4;
– gross regional product in 2007, million rubles—X5;
– fixed assets in the economy (at full cost, at the end of the year), million rubles—X6;
– the volume of shipped goods of own production, jobs performed and services rendered by one's own means in various types of economic activity, million rubles. (mining)—X7;
– the volume of shipped goods of own production, jobs performed and services rendered by one's own means in various types of economic activity, million rubles (manufacturing industries)—X8;
– agricultural output—total, million rubles—X9;
– commissioning of the total area of residential buildings, thousand sq.m.—X10;
– turnover of retail trade, million rubles—X11;
– investments in fixed assets, million rubles—X12.

The presence of a high correlation between the above factors is determined from the condition:

$$r_{xi,xj} > 0.8. \tag{4}$$

According to the current rules, only one of the highly correlated factors is included in the regression equation. In this case, it is expedient to leave one of them, which has the weakest connection with the other factors. The elimination of highly correlated factors is carried out iteratively and the correlation matrix is recalculated at each

Fig. 2 Resulting correlation
matrix

	x1	x2	x9
x1	1		
x2	0,160243	1	
x9	0,619317	-0,08234	1

Fig. 2 Resulting correlation matrix

iteration. As a result of this procedure, there are factors that are not correlated with each other, which will be used in the future modeling.

Figure 2 shows the resulting correlation matrix. It proves that during the multi-collinear test of the selected twelve socio-economic factors, only three can be included in the model:

– average annual number of employed in the economy—X1;
– per capita monetary income (per month), rubles—X2;
– agricultural output—total, million rubles—X9.

These factors will be used in future calculations.

As for the group of factors characterizing the state of human development, they showed a correlation force below the established boundary of 0.8 (r alcohol, drugs = 0.75). Therefore, these factors, as well as the factors that characterize the dynamics of the urban population and youth, are included in the model without adjustments.

The next stage—*the stage of identification and verification of the model*—is the main step in the conduction of modeling. At this stage, the model is constructed, its statistical analysis, parameter estimation are carried out, and the adequacy of the model is verified.

From the point of view of practical implementation, the content of this stage consists in finding the values of parameters $b_0, b_1, \ldots b_n$ of the model (3) on the basis of the available sets of statistical data Y and X.

In this sense, this procedure differs from the traditional algebraic tasks, where the variables Xi ($i=1, 2, \ldots, n$) act as unknowns to be found. In our case, the unknowns are $b_0, b_1, \ldots b_n$, and the values of Y and X are known.

The procedure for finding the values $b_0, b_1, \ldots b_n$ is performed in the module "Data Analysis" of Excel using the "Regression" tool.

With the help of the "Regression" tool, the model parameters are sought from the minimization of its residuals or, more precisely, from the condition for minimizing the function $E(b_0, b)$, which is the sum of the squares of the differences between the actual Y values (taken from the statistics data) and the Y_x values (which characterize the regression equation) calculated for each of the j observations ($j=1, 2, \ldots, m$):

$$E(b_0, b) = \sum_{j=1}^{m} \left(y_j - y_{jx}\right)^2 \rightarrow \min \tag{5}$$

The values which the analyst should consider when estimating the significance of the model parameters and its adequacy:

–the R-square value of 0.84 indicates that the changes in the dependent variable Y (the number of crimes registered under the Article 162 of the Criminal Code of the Russian Federation) basically (by 84%) can be explained by changes in the explanatory variables (factors) included in the model—$X1, X2, X3$ (average annual number of employed in the economy, per capita monetary income, agricultural output). This value proves the adequacy of the model.

The Significance of F. The model is recognized as valid in its entirety if the *value of F* < 0.05, and invalid if the *significance of F* > 0.05. In our calculation, the *significance of F* $= 1.03E-29$, which is less than 0.05. Thus, the conclusion is drawn that the model constructed is generally valid.

p-value. The significance of the found parameters (coefficients) of the model b_1, $b2$, b_3 is estimated from the *p-value*. If the *p-value* is less than 0.05 for the analyzed coefficient, this coefficient is recognized as significance and the corresponding factor (variable) X is considered as one influencing Y. Otherwise, the coefficient is considered insignificant, and the corresponding factor is identified as one not influencing Y.

The data analysis allows us to conclude that only the coefficients of the factors $X1$ and $X3$ are valid since only for them the *p-value* is less than 0.05. Thus, the factor $X2$ is not significant, and its inclusion in the model is impractical. The variable $X2$ is excluded from the consideration, and the calculation is repeated anew.

Upon completion of calculations needed, having ascertained the significance of the found parameters b_1 и b_2, we can construct the desired equation of linear multiple regression:

$$Y = -11.67 + 0.26\,X_1 - 0.0007\,X_2. \tag{6}$$

Let us now consider the meaningful interpretation of the model parameters.

The coefficient $b_1 = 0.26$ shows a slight positive effect on the dynamics of robbery attacks: its value means that the increase of the average annual number of employed in the economy (X_1) by 10 thousand people leads to the robberies increases by an average of 2.6. This result is explained by the fact that the average annual number of employed in the economy is a certain form of reflecting the population size in the regions, with the increase of which crime is growing.

Coefficient $b_2 = -0.0007$ describes the opposite trend: the growth of agricultural output (X2) by 10 billion rubles leads to a reduction in the number of robberies by 7. In other words, the development of agriculture (and in 2014 there was a steady growth of this sector of the economy) has a positive impact on the criminal situation in the Russian regions.

As noted above, the analysis of *p*-values shows that both coefficients are significant.

Similar calculations must be made for all the selected socio-economic factors and crime rates, after which the results should be listed in the table, including the built-up linear models of multiple regressions.

Further, for each of the obtained models, it is required to offer a meaningful interpretation of the dependencies found and the estimated parameters.

6 Analysis of the Results

The study of the final table with the results of calculations allows us to draw the following conclusions.

1. The factors that characterize the state of human development and reflect the consumption of "harmful goods" have the greatest impact on the indicators of crime rate considered in the research. At the same time, a high degree of negative impact of drug use on all indicators of crime rate is observed. The excess of the value of the parameter characterizing the number of patients taken under observation with the established diagnosis of "drug addiction" for the first time in life is two times higher than the parameter associated with the alcoholism factor (for a crime under the Article 111 of the Criminal Code of the Russian Federation) and up to ten times higher (for a crime under the Article 162 of the Criminal Code of the Russian Federation). The proximity of the values of these parameters for indicators of recidivism and juvenile delinquency is of concern. This means that juvenile offenders and recidivists are increasingly willing to replace alcohol with drugs.

    The performed calculations also do not confirm the thesis that high alcohol consumption increases violent crime (in particular, murders); while for property crimes (thefts) the opposite effect is observed. Perhaps, the available statistical data do not fully reflect the real state of crime (as it is known, thefts have high latency). So, it is likely, that we are seeing a change in criminological tendencies. Anyway, the result obtained requires further consideration.

2. Such demographic indicators as the number of young people and the urban population are most closely related to the number of thefts committed (crimes under the Article 158 of the Criminal Code of the Russian Federation) and to the least extent determine violent crimes (Articles 105, 111 of the Criminal Code of the Russian Federation). In addition, there is a positive relationship between the number of young people and the number of recorded crimes committed by persons who have previously committed crimes, which leads to the conclusion that there is a certain decrease in the age of the recidivist. And, although this connection is not so significant, this fact, being established on the national scale, causes some concern.

3. The factor associated with the production of agricultural products does not demonstrate a sustainable impact on most of the analyzed crime indicators, although with some of them (crimes under the Articles 161, 162 of the Criminal Code of the Russian Federation) a negative dependence is observed. However, considering the emerging international situation and the tightening of economic sanctions against the Russian Federation, and also assessing the high agrarian status of the Russian regions, it can be assumed that further development of agricultural production has a great potential for improving the criminal situation in the country.

4. The average per capita money income of the population does not have any impact on crime rates, which corresponds to reality. As it was shown by the criminolog-

ical research carried out in our country and abroad, the growth of social tension in the society and, as a consequence, the increase in the rate of crimes (primarily, violent) are provoked not by the revenues themselves but by their high inequality.

5. The overall adequacy of the results of the study is confirmed by an assessment of the impact of such socioeconomic factor as the average annual number of employed in economics on selected crime rates As we have already noted above, this factor characterizes the population in the regions, with the increase in which crime is increasing. In addition, the ranking of crime rates by the average annual number of employees corresponds to the distribution of the number of crimes committed by the compositions in question. In fact, thefts take the top slot in the group by their number, while murders take the last one.

7 Conclusion

The proposed methodology for analyzing the impact of socio-economic factors on the state of the criminal situation in the regions of the Russian Federation is based on the correct use of a well-known mathematical apparatus and generally available software tools with the help of which the statistical data available to the authors of this work were processed.

It should be noted that the main difficulty in the practical implementation of the proposed methodology is precisely the possible absence of qualitative statistical information that characterizes the socio-economic situation of a given region in dynamics over a period of fifteen to twenty years.

At the same time, even if there is a shortage of necessary data for the entire specified time period, this technology is applicable if we move from a temporary to a spatial form of information presentation, for example, to investigation of the state of the criminal situation in the selected year in the aggregate of administrative and territorial entities that belong to a specific subject of the Russian Federation.

This is the approach we have used in this study to find a set of factors that determine the values of crime rates in the regions of the Russian Federation in 2014.

We'd like especially to emphasize that the expansion of the statistical base will contribute to the further development of the proposed methodology. Moreover, its improvement largely depends on the relevance in practice. If these conditions are met, it is possible to adapt these guidelines for regional and district levels of management of internal affairs bodies.

Analytical materials are stored in the educational and methodical room of the Department of Information Technologies in printed and electronic form.

References

1. Lynch, M., Stretesky, P., Long M.: Defining Crime: A Critique of the Concept and Its Implication, 194 p. Palgrave Macmillan, US (2015)
2. Rosenfeld, R., et al.: The crime drop in comparative perspective: the impact of the economy and imprisonment on American and European burglary rates. In: The International Crime Drop: New Directions in Research. Crime Prevention and Security Management, pp. 200–228. Palgrave Macmillan, UK (2012)
3. Aebi, M.F., et al.: Crime trends in Western Europe according to official statistics from 1990 to 2007. In: The International Crime Drop: New Directions in Research. Crime Prevention and Security Management, pp. 37–75. Palgrave Macmillan, UK (2012)
4. Dahlbäck, Olof: Analyzing Rational Crime—Models and Methods, p. 218. Springer, Netherlands (2003)
5. Rao, C.R., et al.: Linear Models and Generalizations. Springer Series in Statistics, 572 p. Springer, Berlin Heidelberg (2008)
6. Jiang, Jiming: Linear and Generalized Linear Mixed Models and Their Applications, p. 257. Springer, New York (2007)
7. Bingham, N.H., Fry, J.M.: Regression. Linear Models in Statistics, 284 p. Springer, London Limited (2010)
8. Wierzchon, S., Klopotek, M.: Modern Algorithms of Cluster Analysis. Studies in Big Data, 421 p. Springer International Publishing (2018)
9. Murray, A.T., et al.: Exploring spatial patterns of crime using non-hierarchical cluster analysis. In: Crime Modeling and Mapping Using Geospatial Technologies, pp. 181–201. Springer, Netherlands (2013)
10. Groff, E.: Measuring a place's exposure to facilities using geoprocessing models: an illustration using drinking places and crime. In: Crime Modeling and Mapping Using Geospatial Technologies, pp. 269–295. Springer, Netherlands (2013)
11. Cramer, D., Brown, A.A., Hu, G.: Predicting 911 calls using spatial analysis. In: Lee, R. (eds.) Software Engineering Research, Management and Applications 2011. Studies in Computational Intelligence, vol. 377. Springer, Berlin, Heidelberg
12. United Interdepartmental Statistical Information System. https://www.fedstat.ru/indicators. Accessed 21 Dec 2017

The Remote Approach of Distribution of Objects Withdrawn from Circulation: Means, Legislation Issues, Solutions

Yuri V. Gavrilin, Nikolay V. Pavlichenko and Maria A. Vasilyeva

Abstract Penetration of criminal activity into the communication sphere, including the Internet. Use of the Internet possibilities for the illegal sale of items prohibited for turnover, such as narcotic drugs, psychotropic substances, weapons, pornographic images (including child ones), etc. Distribution of crypto-currencies as means of payment while committing illegal acts. The purpose of this chapter is to analyze the mechanism of the remote sale of objects withdrawn from turnover, to detect conditions that contribute to this illegal activity, to justify legal means of its constraints, as well as criminalistics technologies of its identification, termination, and investigation. The following recommendations are able to increase the efficiency of the prevention of illegal traffic of prohibited items: imposing on the Internet providers obligations to store user correspondence, transmitted audio and video images and other content; proper identification of users; restrictions on the use of anonymizers, as well as on the use of Internet resources by persons previously convicted of certain crimes; making the owners of the Internet resources responsible for posting prohibited information by users; control of transactions using the crypto-currency, and close international cooperation in law enforcement. The above list of measures aimed at limiting the use of the Internet in unlawful activities will provide a minimum of inconvenience to bona fide users, but it can have a deterrent effect on the distribution of prohibited items, which ultimately will have a positive impact on the legal protection of citizens.

Keywords Commission of crimes · A way of selling items · Remote mode
Crime prevention

Y. V. Gavrilin · N. V. Pavlichenko (✉) · M. A. Vasilyeva
Academy of Management of the Ministry of Internal Affairs of Russia, Moscow 125993, Russia
e-mail: aldan1973@rambler.ru

Y. V. Gavrilin
e-mail: yuriy902@gmail.com

M. A. Vasilyeva
e-mail: mvd-ecology@mail.ru

© Springer Nature Switzerland AG 2019
A. G. Kravets (ed.), *Big Data-driven World: Legislation Issues and Control Technologies*, Studies in Systems, Decision and Control 181,
https://doi.org/10.1007/978-3-030-01358-5_8

1   Introduction

Active development of digital economy, the expansion of use of information and communication technologies, increase of their availability, growing number of users of information resources of the Internet, emergence of new tools, techniques and services have exerted a powerful influence on virtually all spheres of activity of citizens, organizations, states and the entire world community [1]. According to the International Telecommunication Union, the number of the Internet users now exceeds 3.5 billion people, most of whom (2.5 billion people) live in the developing countries. While in the developed countries the Internet penetration rate averages at 81%, which is more than twice the level of the developing countries [2]. According to the Fund "Public Opinion", the level of the Internet penetration in Russia, as of May 2017, was 70% [3].

The existing level of development of information and communication technologies naturally has an impact on the criminal sphere. A substantial number of "traditional" crimes, such as threats of murder, abetment of suicide, copyright infringement, theft, fraud, extortion, illegal purchase and sale of weapons, drugs, crimes of extremist and terrorist nature, illegal distribution of pornographic materials are committed remotely (i.e. without direct contact with the victim [4]). The remote method of crime commission presupposes that the interaction of the offender with the victim or third parties is mediated using information resources of the Internet, primarily such as social networks, e-mail, instant messaging services, etc.

On the basis of generally recognized acts of the international law, such as the Single Convention on Narcotic Drugs (concluded in New York on March 30, 1961), the Convention on the Means of Prohibiting and Preventing the Illicit Import, Export and Transfer of Ownership of Cultural Property (concluded in Paris on 14.11.1970), the Convention on Psychotropic Substances (concluded in Vienna on February 21, 1971), the Convention on the Prohibition of the Development, Production and Stockpiling of Bacteriological (Biological) and Toxin Weapons and on Their Destruction (concluded on 16.12.1971), the Convention on the Prohibition of the Development, Production, Stockpiling and Use of Chemical Weapons and on Their Destruction (concluded in Paris on 13.01.1993), and others, the legislation of the absolute majority of states prohibits or restricts the turnover of narcotic drugs, psychotropic substances, weapons, stolen cultural property and other objects prohibited for free circulation (hereinafter referred to as items withdrawn from circulation).

2   The Technological Basis of the Remote Method of Distribution of Objects Withdrawn from Circulation

The wide spread of the remote method of committing crimes connected with the illegal sale of items withdrawn from circulation is conditioned by the following factors:

- ensuring anonymity in communication, eliminating the need for a direct salesperson and acquirer;
- the possibility of using cryptocurrency in mutual settlements;
- the accessibility of the Internet, the ease of disseminating information to a wide audience;
- the possibility of illegally distributing items withdrawn from circulation in the territory of other regions and states;
- securing the concealment of information about the subjects of the crime by using e-mail addresses and subscriber numbers for accessing the Internet network registered in other regions or in other states by fictitious persons ("dummies") [5].

Withholding of information about the identity of the distributors of objects withdrawn from turnover is also provided by the use of e-mail servers located in the territories of foreign states, with which there are no international treaties on legal assistance, as well as by the use of the TOR browsers [6]. The latter is a package of special software designed for anonymous Internet search, concealing the configuration of the user equipment and its location. Confidentiality is provided by routing and encryption of information received/transmitted over a distributed network of servers located in different parts of the world.

These circumstances are decisive when criminals use a remote method of selling items that have been withdrawn from circulation.

The considered method of selling items withdrawn from turnover can be carried out on the basis of various Internet technologies, in particular:

1. Creation of specialized sites which offer prohibited items through the use of the servers located outside the jurisdiction of the state with a targeted audience. The presence of such sites is easily identified by specifying in the search line of any Internet browser (Google, Yahoo, AOL, Yandex, etc.) the name of an item withdrawn from circulation. These sites are essentially illegal Internet shops, through which you can not only purchase items withdrawn from turnover but also get detailed instructions on their application, methods of payment and delivery, as well as other information related to these information objects. In order to increase the attractiveness of the goods sold, these sites make use of such marketing instruments as discounts, actions "three items for the price of two", etc.
2. Use of the opportunities of social networks for illegal sales of objects withdrawn from the turnover. In special literature social network is defined as the Internet service that enables users:

  - to create open or partially open users' profiles containing, for example, date of birth, marital status, education, hobbies, profession, etc.;
  - to create a list of users with whom they have social relations (friendship, common interests, professional connections, etc.);
  - to view the relationship between users (the ability to see friends of your friends) [7].

The above list of opportunities of social networks can be supplemented with such functions as:

- search for the needed contacts to establish relationships (friendship, commercial, professional, interests, etc.);
- ensuring communication of users by sending to each other text messages and multimedia content (graphics, photos, and videos, audio recordings, etc.) as well as verbal communication;
- stimulation of the achievement of certain goals of visiting this Internet resource (searching for acquaintances, keeping a diary (blog), receiving and disseminating information, confirming it or refuting it, etc.);
- changing the indicators that determine the personal status of the user in the social network (the number of page views, the approval of the posted message or other content, the number of friends, subscribers, etc.);
- gaining practical benefits from participating in a social network (increasing personal self-esteem, finding a job, obtaining socially useful links, positive evaluation by other persons, creating a family, etc.) [8].

The wide spread of mobile communication, wireless high-speed data transmission for mobile devices, as well as reducing the cost of such services, gave a significant impetus to the development of social networks. In the end, the audience of social networks demonstrates the high values of the percent increase. Thus, according to data cited by the Russian Business magazine, the audience of the most popular social network in the world "Facebook" in April 2017 was 1968 million users. In the second place, there is a service for instant messaging WhatsApp—1200 million users; the third position is taken by the video hosting YouTube—1000 million users [9]. The most popular social networks in the Russian Federation (according to the Fund "Public Opinion") are "VKontakte" ("In Contact")—61%, "Odnoklassniki" ("Classmates")—57%, Facebook—16% [10].

Use of a particular Internet technology when making remote sales of items withdrawn from turnover may be carried out in the following areas:

- dissemination of information on the availability of acquisition of certain objects withdrawn from turnover, their properties (characteristics). Such distribution may be performed for an indefinite number of persons (for example while creating Internet sites) or for a specific (target) audience (for example while creating a group in the social network);
- establishing of a direct contact between distributor and purchaser of the objects withdrawn from circulation, matching of price and name; and in some cases, the range of the acquired objects through correspondence or telephone conversations;
- informing the customer about methods and order of payment (use of e-wallets, crypto-currencies, transfer of funds to the account of a mobile phone via self-service terminals, other ways of payment where identification of the payer is not possible);
- informing the buyer about the place and time of receiving the item withdrawn from turnover in a certain hiding place, about which the buyer is informed after the purchaser has received the payment from him. It is usually done by sending the buyer the geolocation data, photographic images or verbal description of the hiding place;

– ensuring information exchange between participants of a criminal group engaged in illegal sales of items withdrawn from circulation in accordance with their role in the criminal hierarchy and criminal specialization [11].

The above suggests that modern criminality has actively adopted the latest achievements of the digital economy and information technologies, thereby ensuring, on the one hand, the expansion of the criminal market of items withdrawn from turnover and, on the other hand, hiding the identity of criminals. In this connection, considering the fact, that there are no frontiers for crimes committed remotely, they can be effectively controlled only if the actions of all states are coordinated [12].

3 The Legal Basis for Combating the Proliferation of Items Withdrawn from Circulation

The basic international legal instrument to ensure the international cooperation in the fight against crimes committed using information and communication technologies (including remote sales of items withdrawn from circulation) is the Convention on Crime in the Field of Computer Information Systems (ETS No. 185) which was concluded in Budapest on 23.11.2001.

This document provides a classification of crimes committed using information and communication technologies, regulates the procedural aspects of obtaining evidence on electronic media, and establishes the principles of international cooperation in this field.

At the same time, it should be noted that the list of criminal acts contained in the Convention does not comprise such an act as the illegal sale of objects withdrawn from circulation, committed by a remote method. At the same time, in the Russian Federation, there is criminal liability for the illegal sale or transfer of narcotic drugs, psychotropic substances and their analogues using electronic or information and telecommunication networks (including the Internet) (paragraph "b" part 2, Article 228.1 of the Criminal Code), distribution, public demonstration or advertising of pornographic materials or objects, materials or items with pornographic images of minors using information and telecommunication networks (including the Internet) (paragraph "b" part 3 of the Article 242, paragraph "g" of part 2 of the Article 242.1 of the Criminal Code of the Russian Federation).

At the same time, the Convention establishes certain procedural powers of the competent authorities of the Contracting States, in particular [13]:

– to require the preservation of certain computer data, that identify the service providers and the way in which the information was transferred, as well as the confidentiality of these procedures;
– to require the persons to provide information at their disposal, and from service providers - information about their subscribers;
– to gain access to computer systems, their parts and data stored in them during the search and seizure;

- to produce and keep copies of relevant computer data;
- to block access to certain computer data;
- to require providing the necessary information from any person with knowledge of the functioning of the relevant computer system or the measures taken to protect the computer data stored there;
- to collect or record data on information flows transmitted through a computer system or oblige service providers to perform specified actions within their technical capabilities.

4   Russian Experience in the Legal Regulation of the Use of Information and Communication Technologies

Even though the Convention had not been ratified by the Russian Federation, it had a significant impact on the development of legislation in the sphere of combatting crimes committed using information and communication technologies, in particular:

- the Federal Laws No. 143-FL of 28.07.2012 and No. 375-FL of 06.07.2016 regulate the procedure of withdrawal of electronic media during search and seizure;
- the Federal Law No. 139-FL of 28.07.2012 establishes the unified register of domain names, the indication of pages of the Internet sites containing the prohibited information so that to limit access to the Internet sites containing information dissemination of which is prohibited in the Russian Federation [14];
- the Federal Law No. 110-FL of 05.05.2014 strengthens control over payments made online including restrictions for anonymous payments;
- the Federal Law No. 97-FL of 05.05.2014 stipulates the responsibility of the communication operator, which is charged with the identification of users of communication services for data transmission and provision of access to the Internet and the terminal equipment used by them.

In addition, the organizer of the dissemination of information via the Internet has the responsibility to store information about the receipt, transmission, delivery and/or processing of voice information, written text, images, sounds or other electronic messages of Internet users and information about these users for six months from the moment of the end of the implementation of such actions, and provide the said information to the authorized state bodies.

By the Federal Law No. 374-FL of 06.07.2016, the organizer of information dissemination on the Internet has the following responsibilities:

- to keep the above-mentioned information within one year from the date of completion of implementation of such actions;
- to keep the text messages of the Internet users, voice information, images, sounds, video, and other electronic messages of the Internet users for up to six months after the end of their reception, transmission, delivery and (or) processing;
- to provide the above-mentioned information to authorized state bodies;

- to provide the authorities with the information necessary for decoding the messages [15], when using the additional encoding of electronic messages and (or) providing the Internet users with the possible additional encoding of electronic messages for the reception, transmission, delivery and (or) processing electronic communications of the Internet users;
- the Federal Law No. 375-FL of 06.07.2016 regulates the procedure for seizure of electronic communications and other messages transmitted via telecommunication networks;
- the Federal Law No. 276-FL of 29.07.2017 16 establishes a ban on the use of anonymization programs when accessing information resources, access to which is restricted in the territory of the Russian Federation;
- the Federal Law No. 241-FL of 29.07.2017 obliges the organizer of instant messaging to ensure the transmission of electronic messages only to those Internet users who have passed the appropriate identification procedure (using the subscriber number, based on the relevant agreement). In addition, the messengers must: within 24 h from the moment of receipt of the corresponding requirement of the authorized body to limit the possibility of the user performing the messaging service; to provide a technical possibility for users of the instant messaging service to refuse from receiving electronic messages from other users; to store identification information about the subscriber number only in the territory of the Russian Federation.

Thus, the current Russian legislation is brought into compliance with the European standards specified by the European Convention of 2001.

However, are there enough measures taken to reduce the criminal potential of the Internet in terms of the remote distribution of prohibited objects? It seems that an affirmative answer to this question is premature [15]. The Internet still has a large number of proposals for the illicit sale of items withdrawn from circulation.

5 Conclusion

As additional measures aimed at reducing the volume of the illegal market for such facilities, the following is proposed.

1. To oblige the Internet search systems operators not to permit issuance of information about the index of the page of the site on the Internet, which allows access to information containing proposals for the acquisition of items withdrawn from circulation. In other words, the operators of search systems (Yandex, Google, Yahoo, Bing, MSNSearch, Rambler, Mail.ru, etc.) should provide by themselves monitoring of information resources links to which are given on the search queries of users. Today technologies for semantic analysis of social media content and Internet sites are developed and are actively used in practice when conducting marketing research, that indicates the availability of technical opportunities to implement such requirements.

2. To oblige the Internet search systems operators not to give information on search queries containing certain key phrases (like "to buy heroin", "synthesis of amphetamine", "to get spice", etc.). Currently, major search systems have special services (Google Analytics, Yandex, Wordstat., etc.) that allow you to analyze information on statistics and content of search queries, which also indicates the technical feasibility of such requirements.

3. In connection with the widespread use of crypto-currencies ("virtual currencies") allowing making anonymous transactions in the absence of any control by the competent authorities, it is necessary to consider the issue of legal regulation of their turnover with an obligatory procedure for identifying parties to the transaction, where means of payment is crypto-currency.

4. The introduction of an additional punishment into the criminal legislation in the form of a ban of up to 10 years of using the information and telecommunication network of the Internet for persons found guilty of committing crimes related to the illegal sale of items withdrawn from the circulation using the Internet.

5. Expanding international cooperation in combatting crimes committed using information and communication technologies, both by concluding separate (bilateral and multilateral) agreements, and by adopting a universal UN Convention on International Information Security aimed at harmonization of national legislation and containing uniform approaches to combatting such crimes [16].

The above list of measures aimed at limiting the use of the Internet in unlawful activities will provide a minimum of inconvenience to bona fide users, but it can have a deterrent effect on the distribution of prohibited items, which ultimately will have a positive effect on the legal protection of citizens.

References

1. Buono, L.: Fighting cybercrime through prevention, outreach and awareness raising. article. J. Acad. Eur. Law (ERA). **15**(1), 1–8 (2014). https://link.springer.com/article/10.1007/s12027-0 14-0333-4. Accessed 01 Dec 2017
2. Press release of International Telecommunication Union. Geneva, 22.07.2016. www.itu.int/ne t/pressoffice/press_releases/2016/pdf/30-ru.pdf Accessed 23 Aug 2017
3. http://fom.ru/SMI-i-internet/13585 Accessed 24 Jul 2017
4. More detailed see: Was jeder über Viren, vol. 10. Spam und Datenklau wissen sollte, Heidelberg, Dec 2015
5. Armstrong, C.J.: Mastering computer forensics. Article. J. Secur. Educ. Crit. Infrastruct. 151–158. http://link.springer.com/chapter/10.1007/978-0-387-35694-5_14. Accessed 01 Dec 2017
6. Subrahmanian, V.S., Ovelgonne, M., Dumitras, T., Prakash, B.A.: The Global Cyber-Vulnerability Report, XII, 296 p. (2015)
7. Boyd D., Ellison N.: Social network sites: definition, history, and scholarship. J. Comput.-Mediated Commun. **T.13**(1). 220 (2007)
8. Solovyov, V.S.: Criminogenic potential of the social segment of the internet: methodology assessment and neutralization marks. Monograph, pp. 8–9. Krasnodar (2016)
9. https://rosbj.ru/2017/05/13/1250-most-popular-social-networks-2017. Accessed 23 Aug 2017
10. https://fom.ru/SMI-i-internet/12495. Accessed 24 Aug 2017

11. Kshetri, N.: Information and communications technologies, cyberattacks, and strategic asymmetry. In: The Global Cybercrime Industry. Springer, Berlin, Heidelberg (2010)
12. Buono, L.: Fighting cybercrime through prevention, outreach and awareness raising. J. Acad. Eur. Law (ERA). **15**(1), 1–8 (2014). http://link.springer.com/article/10.1007/s12027-014-0333-4 Accessed 01 Dec 2017
13. Synodinou, T.-E.: In: Regulation and Enforcement, 1st edn., XX. 433 p. (2017). ISBN 978-3-319-64955-9
14. A list of such information is set by part 2 of article 5 of the Federal Law of 29.12.2010 №. 436-FL "On Protection of Children from Information Harmful to their Health and Development", paragraph 6 of the article 10 of the Federal Law of 27.07.2006 №. 149-FL "On Information, Information Technologies, and Protection of Information"
15. von Solms, B.: Securing the internet: fact or fiction? Article. http://link.springer.com/chapter/10.1007/978-3-642-19228-9_1. Accessed 01 Dec 2017
16. Brinkhoff, S.: Big data mining by the dutch police: criteria for a future method of investigation. Article. Eur. J. Secur. Res. **2**(1), 57–69 (2017). http://link.springer.com/article/10.1007/s41125-017-0012-x. Accessed 01 Dec 2017

Remote Investigative Actions as the Evidentiary Information Management System

Evgeny Kravets, Svyatoslav Birukov and Mikhail Pavlik

Abstract The urgent character of the research is associated with the necessity to eliminate a systemic noncompliance with the level of up-to-date information and communication technologies integration with the needs of criminal investigation activities. This qualitative characteristic reduces the efficiency of law enforcement. The further disregard of the ever-increasing innovation deficiency can lead to irreversible negative consequences.

Keywords Investigative action · Information
Information and communication technologies · Video conferencing
Evidence information

1 Introduction

The origin and movement of information in the criminal investigation, in particular when conducting investigative actions, is a highly complicated and quite controversial process. To comprehend its regularities it is reasonable to refer to philosophy, psychology, computer science, and other natural as well as humanitarian sciences. To be able to use such information it is necessary to evaluate it, determine if it is clear to those individuals who analyze it, establish its reliability, admissibility, relevance to a criminal event, and only then resolve an issue of its further use.

E. Kravets (✉)
Volgograd Academy of the Russian Ministry of Internal
Affairs, Volgograd, 130 Istoricheskaya str., Volgograd 400089, Russia
e-mail: 80kravez@gmail.com

S. Birukov
Volgograd State University, 100 Universitetskiy av., Volgograd 400062, Russia
e-mail: bir.slav@yandex.ru

M. Pavlik
Pushkin Leningrad State University, Sankt-Peterburg 196605, Russia

© Springer Nature Switzerland AG 2019
A. G. Kravets (ed.), *Big Data-driven World: Legislation Issues and Control Technologies*, Studies in Systems, Decision and Control 181,
https://doi.org/10.1007/978-3-030-01358-5_9

The analysis of a semantic component implies detailing the conceptual content of data and relations between them, in other words, understanding the meaning and significance of the obtained evidentiary information. In the opinion of Samygin [1], prior to repelling the interrogatee's statement the investigator should perceive his testimony and understand as accurately as possible its meaning for a particular case. The evaluation of a semantic aspect of information during an investigative action implies forming certain signals related to a crime under investigation from all the incoming information signals.

A pragmatic aspect of evidentiary information is primarily connected with the importance of using this testimony while generating a solution to the problem brought before the court. In general, the value of information can be measured by the degree to which a goal is achieved [2]. Depending on the purpose of the same information, it can be either useful or useless [3].

It is a variety of information signals which is necessary and sufficient to achieve a goal by selecting from a whole existing variety that gives an idea of the pragmatic value of evidentiary information. However, the category of usefulness and valuableness of information can hardly be unambiguously evaluated since in each particular case data seeming excessive in a certain combination may turn out to be valuable and useful under the changed conditions.

It is this context in which any investigative action "serves as a specific set of cognitive techniques of finding out evidentiary information of a certain type [4]." According to numerous research works of the last few years, up to three-quarters of the whole array of evidentiary information are contained in the testimony of criminal proceedings' participants [5]. That is why the analysis of the information nature of investigative actions must be mainly based on the study of interrogation/interview and face-to-face confrontation. Moreover, when conducting lineups an identifying person also testifies and even is generally warned of liability for giving knowingly false testimony or refusing to testify.

2 Factors Stipulating the Obtaining and Amount of Information to be Used in a Criminal Investigation

Factors stipulating the obtaining and amount of information to be used in a criminal investigation are classified into subjective and objective ones. Objective factors imply peculiarities of investigation agencies' powers to carry out activities related to criminal procedure and peculiarities of the criminal-law characteristic of a crime. Subjective factors include the amount of initial information characterizing a participant of criminal proceedings (in the beginning of investigation), some extra time for preparing for an investigative action, the qualification level of the investigator conducting an interrogation/interview, a degree of staffing of a corresponding law enforcement agency, participation of a translator when obtaining data from foreign nationals, technical equipment, the level of using special knowledge while conducting an investigative action, and practice of interaction between an individual conducting an investigation and operational officers.

The evidentiary information obtained by the investigator cannot exist beyond the bounds of an information signal that is a particular physical process (alteration) conveying information about an object, phenomenon, or event, as the model of an object, phenomenon, or event [6]. The concept of a signal of information about an event under investigation corresponds to reflection in a modified form. The signal is a result of the mutual influence of at least two processes or structures: a human being and an environment, a subject and an object, and serves as a connecting link between them [7]. It comes from the event or action being perceived and is directly connected with them. The signal exists independently within a certain organized system. It can be recorded, transmitted over large distances, and exist for a long time in the same form.

In some conditions, there is an unambiguous connection or compatibility within the limits determined by the description accuracy between the event and the signal [8]. The system of information signals consists of the following elements:

– a source of information (the process of information arising);
– a transmitter of information (a participant of an investigation action);
– an addressee (an individual conducting an investigation) [7].

Apart from the content of the signal, its form also plays an important role. It is served by the type of information's existence. Information cannot be transmitted and processed without a form. It is an information code involving various means of communication that is a form of expressing the signal.

An investigative action is a process of joint information and mental activity, mutual influence of the investigator and the interrogatee/interviewee on the reconstruction of circumstances and facts which are significant for investigating a case. The information circulating during an investigative action and being an object of exchange for its participants is quite diversified.

The quality of information deteriorates if it is frequently transmitted. The reduction of information value of the obtained indications can turn out to be the result of errors in perception and defects of a participant's sensory apparatus, his inability to reproduce what was perceived, interpretation errors made by the investigator, an improper record or judges' misunderstanding of the interrogation/interview record content. It's obviously wrong to equate the quantity of information contained in the matter under study to the ideal model and an information message about it. Interrogatees/interviewees describe their recollections of the event which have already gone through consciousness and not what they perceived. Along with this, it is truly important to determine how accurately the reality is reflected in the ready-to-use information, in other words, how correctly it is entered, converted, and conveyed by the interrogatee/interviewee, perceived and recorded by the investigator. A video recording of an investigative action may appear to be a truly valuable means of compensation for recording defects.

As a rule, information is partially lost every time it is transmitted. Yet people speaking the same language understand each other. Such a loss of information is inevitable during an investigation in general and interrogation/interview in particular. However, it can be made up of new sources due to obtaining other data and combining a series of fractional stages of information transmission [9]. The main role here belongs to taking testimony from the first person who is an eyewitness of the very crime or significant circumstances being a means of detecting its elements as well as to using direct evidence preventing the loss of information and the probability of its distortion that is the principle of directness [10].

The participant's unawareness of the language of court proceedings is an additional factor of evidentiary information distortion. A translator is engaged to assist in the conduct of an interrogation/interview of foreign nationals as well as those individuals who don't speak Russian. According to Article 59 of the Code of Criminal Procedure of the Russian Federation, a translator is an individual being engaged in criminal proceedings and being fluent in the language needed for translation.

The process of conducting an investigative action always implies communication and a tense intellectual mutual influence of the involved parties. It results not only in the recording of the information obtained from the interrogatee/interviewee but also a consequence of the investigator's comprehensive activity. The process of communication is directly connected to the mnemonic abilities of the interrogatee/interviewee. Taking the most objective, complete, and comprehensive testimony predetermines his memory and perception.

Based on the said above, it can be concluded that any moment of mental activity of the investigation's participants under analysis reveals specific aspects of the psyche. There are volitional, emotional, and cognitive processes which can always be distinguished in it. Prompting certain actions and deeds they influence the further activity of the personality [11].

Regularities of information movement when conducting lineups are similar to interrogation/interview and face-to-face confrontation. It is mainly related to the fact that testimony of an identifying person about specific features of a person being identified precedes this investigative action. Of course, it should be taken into consideration that mental reactions of an identifying person are poorer than of the one being identified. They mostly come to associating the seen individual with recollections of him. Besides, in terms of variety of applicable tactical techniques a lineup is, tentatively speaking, a passive investigative action. In fact, while conducting it the investigator's activity apparently comes only to the observation of the very process of identification [12].

During the examination a human body is a source of evidentiary information. That is why the mental activity of a person being examined may be analyzed only in terms of counteraction to the efficient conduct of an investigative action when particular traces, objects, and distinguishing marks are knowingly concealed or procedural powers restricting possibilities of additional recording of evidence are used (refusal to give consent to photo or video recording).

3 Scientific and Technological Means for Information Recording and Seizing

The use of scientific and technological means for information recording and seizing plays a primary role in the process of evidence collection [13]. And not even because there can be no search and preservation in those situations when information carriers are evident. What is more important, recording (reflecting or fixing) of evidence is strictly inherent in criminal procedure and is its mandatory attribute. The role of the officer responsible for a criminal case is especially obvious here. It implies comprehending verbal speech and putting it into the system of non-verbal information signals (mimicry, gestures, pauses, etc.), selecting the most essential aspects, verifying the reproduction correctness by clarifying and test questions, determining the identity of terms to be used, and getting the whole picture of the individual's abilities of perception, memorization, etc.

Nowadays, information and communication technologies (ICTs) are still used by law enforcement agencies in their professional activity. For instance, the information and analytical support system of the Ministry of Interior of Russia are widely used for holding meetings. It contains a lot of information segments providing access to the vast database including operational, reference, and forensic centralized records. The practice of equipping particular subdivisions with local computer networks is also widespread. It facilitates the circulation of proprietary information within them. Moreover, electronic mail, IP-telephony, and facsimile communication are currently used almost everywhere.

It is necessary to point out that not all the above-mentioned ways of using ICTs may be characterized as forensic and technical because the information obtained via these channels belongs to the category of orientation. Nowadays, ICTs aren't used for detecting and recording of evidence. It is known that local networks consisting of two series-connected personal computers have been repeatedly used in practice while conducting lineups under the conditions excluding visual observation of an identifying person. However, the total number of such investigative actions is so insignificant that it teeters on the brink of statistical error.

In current conditions, computer software is supposed to be a key component of the process of recording evidentiary information while conducting investigative actions. A great number of software systems and packages used in law enforcement agencies can be systematized as follows:

- training systems;
- expert and consulting systems;
- information and reference systems;
- calculation and analytical systems;
- management systems and automated workstations;
- image processing systems.

In forensic science the following ways of how to use electronic computing machines are distinguished:

– obtaining data about various objects, processes, and phenomena, automation of
  the initial processing of these data;
– using electronic computing machines and automatic devices in order to obtain
  derived parameters from fixed initial information and for the purpose of complex
  information processing;
– automation of information scanning and coding procedures;
– studying mathematical models of the process of proving;
– computer pattern recognition [14].

The circulation of information during an investigative action is mainly connected
with the effective use of special knowledge and technical means. The successful
outcome of an investigative action and its information saturation directly depend on
possibilities of their application in a criminal investigation by investigators (inquiry
officers), specialists, and courts [15]. At the present stage of the Russian society
development, the role of science and technology in criminal investigation appears to
be especially important. To a certain extent, it is due to the general accessibility of
these achievements.

As a result, at the moment the above-mentioned facts count in favor of the adoption
of ICTs as the means of forensic technical support for arrangement and conduct of
investigative and procedural actions. Accordingly, possibilities and prospects of the
use of e-mail, IP-telephony, and video conferencing in a criminal investigation should
be considered.

4   Possibilities and Prospects of the Use of ICTs
in a Criminal Investigation

As for the prospects of using electronic mail for the needs of a criminal investigation,
it's almost impossible to imagine the up-to-date information exchange including personal and business correspondence without this service. There can be distinguished
the following possibilities of using e-mail:

– for making the investigation's participants aware of criminal case files to the extent
  required;
– for forwarding statements, complaints, and petitions to the investigator or the
  head of an investigative agency and further notification of the results of their
  consideration;
– for providing the participants' appearance before the investigator.

This initiative may be implemented according to the following algorithm: the agency
activates an isolated server, and then on its basis within an autonomous domain there
will be the possibility to create mailboxes for staff members as well as for particular
persons involved in criminal cases. Functional peculiarities of such accounts must
be differentiated depending on the status of a user. Along with that, it is necessary
to take account of the requirements of agency-level orders related to the circulation

of restricted-access information as well as protection from mass advertising and malicious software mailing.

Mailboxes of the involved persons (victims, suspects, defendants, defenders) can be temporarily created before the court decision on a particular criminal case comes into force. Official accounts here are obviously more preferable than personal ones. It is due to their integration into a specific software environment, impossibility to block documents sent by the investigator, and other settings. In the author's opinion, it would be perfect to integrate such a service into an operating environment of the state services portal. It would allow the participants who are not law enforcement officers to get access to their mailboxes, for example, from post offices.

The involved persons may be made aware of criminal case files in the course of the investigation as well as upon its completion in the manner stipulated by Articles 215-217 of the Code of Criminal Procedure of the Russian Federation. In the first case, apart from records of investigative actions in which the individual took part the investigator's orders affecting his interests, reports of forensic experts, and some other materials are also subject to obligatory submission. Besides, when a new defender enters into a case at his request he can be provided with the previously collected materials related to his defendant.

While attaching documents subject to submission to the file the investigator may digitize them by scanning and send to the earlier created e-mail address along with the standard explanation of the order of familiarization and the notification of the time to visit. If certain requirements of the law are revised after a while in order to simplify the procedure the investigator will probably ask to fill out an electronic form for familiarization or send a scanned record for printing out and attaching to the file.

It should also be mentioned that there is a possibility to familiarize the suspect or the accused with the materials forwarded to the court at the investigator's request. If it is arranged via e-mail the personal attendance of these participants will not be necessary. And it will not prejudice their interests. It may become an additional factor in saving because of the funds provided for their transportation to the court.

The same order may be reasonably applied to forwarding and considering complaints, petitions, and other statements addressed to the investigator. The remote receiving of applications will significantly accelerate the resolution process. It is especially helpful for the accused held in custody since their transportation from the pretrial detention center is carried out only several times a week and only at the prior request of an authorized person. And the access to a specialized e-mail may be provided from a penal institution.

The author's proposal to use e-mail for providing the participants' appearance before the investigator is absolutely obvious. According to the current Instruction [16] even reports of crimes may be sent via e-mail. That's why, in the author's opinion, there are no restrictions on sending subpoenas to specialized accounts. Taking into consideration that this mission is urgent addresses may be attached to a number of a cell phone or any other individual means of communication by a notification option.

IP-telephony, as it was said before, may be used when the investigator needs the specialist's consultation on how to raise questions to the expert as well as to ensure the suspect's or the accused's right to be defended. The main advantage of this method over traditional communication is in its low cost. Besides, a normal telephone cannot provide simultaneous communication between several users or communication network services.

When analyzing peculiarities of video conferencing it is reasonable to take account of forensic tactics development in so far as it relates to the vision of an investigative action structure. In the most widespread opinion [17, 18], a traditional structure of an investigative action involves three stages—preparatory, operational, and concluding. The content of each stage varies depending on the specific character of an investigative action. From the author's point of view, remote investigative actions don't require revision of these elements. It should be recalled that among the priority objects of this research there are such investigative actions as interrogation/interview, face to face confrontation, lineup, and examination.

There can be preconditions of exceptionally tactical character for refusal to conduct an investigative action. To a lesser extent, it is related to conflict situations during interrogation/interview and, to a greater extent, to making a decision about the conduct of face-to-face confrontation. As for the confrontation, there are instances when the investigator predicts a threat of losing evidence as a result of such an investigative action. Here it is necessary to take account of the correlation of psycho-emotional characteristics of interrogatees/interviewees, different types of dependence between them, and other factors.

While preparing for the conduct of an investigative action the following parameters should be determined: the time when an investigative action starts, a place of its conduct, its participants (a method of providing their appearance), the content of an investigative action (as a rule, along with a plan), and a set of technical means to be used. All these issues may be resolved on the basis of a thorough study of criminal case files.

References

1. Samygin, L.D.: Criminal investigation as the system of activities, p. 73. Moscow (1989)
2. Astafurova, O.A., Salnikova, N.A., Lopukhov, N.V.: Means of Computer Modeling of Microwave Devices and Numerical Methods as Their Base Communications in Computer and Information Science, vol. 466, pp. 630–642. CCIS (2014)
3. Seravin, L.N.: Laws of Information and Its Role in Human Society, p. 24. Saint-Petersburg (1997)
4. Shafer, S.A.: The Essence and Ways of Collecting Evidence in Soviet Criminal Procedure, p. 41. Moscow (1972)
5. Streltsov, L.: The system of cybersecurity in Ukraine: principles, actors, challenges, accomplishments. Eur. J. Secur. Res. **2**, 147 (2017). https://doi.org/10.1007/s41125-017-0020-x
6. Nan, C., Sansavini, G.: A quantitative method for assessing resilience of interdependent infrastructures. Reliab. Eng. Syst. Saf. **157**, 35–53 (2017)
7. Kröger, W.: Securing the operation of socially critical systems from an engineering perspective: new challenges, enhanced tools and novel concepts. Eur. J. Secur. Res. **2**, 39 (2017). https://doi.org/10.1007/s41125-017-0013-9

8. Scavuzzo, M., Nitto, E.D., Ardagna, D.: Experiences and challenges in building a data intensive system for data migration. Empirical Softw. Eng. **23**, 52 (2018). https://doi.org/10.1007/s106 64-017-9503-7

9. Verma, T., Ellens, W., Kooij, R.: Context-independent centrality measures underestimate the vulnerability of power grids. Int. J. Crit. Infrastruct. **11**(1), 62–81 (2015)

10. Ahmadi, Z., Kramer, S.: Knowl Modeling recurring concepts in data streams: a graph-based framework. Inf. Syst. **55**, 15 (2018). https://doi.org/10.1007/s10115-017-1070-0

11. Das, S., Banerjee, M., Chaudhuri, A.: An improved video key-frame extraction algorithm leads to video watermarking. Int. J. Inf. Tecnol. **10**, 21 (2018). https://doi.org/10.1007/s41870-017-0054-3

12. Brinkhoff, S.: Big data data mining by the dutch police: criteria for a future method of investigation. Eur. J. Secur. Res. **2**, 57 (2017). https://doi.org/10.1007/s41125-017-0012-x

13. Ebad, S.A.: Regulatory rules for security requirements of financial information systems: attempt to formalize. Eur. J. Secur. Res. **2**, 97 (2017). https://doi.org/10.1007/s41125-017-0017-5

14. Azizi, A., Ghafoorpoor Yazdi, P., Hashemipour, M.: Interactive design of storage unit utilizing virtual reality and ergonomic framework for production optimization in manufacturing industry. Int. J. Interact. Des. Manuf. (2018). https://doi.org/10.1007/s12008-018-0501-9

15. Sorell, T.: Online grooming and preventive justice. Crim. Law Philos. **11**, 705 (2017). https://doi.org/10.1007/s11572-016-9401-x

16. Kravets, E.: Cognitive activity efficiency factors during investigative actions, performed using information and communication technologies. In: Kravets, E., et al. (eds.) Communications in Computer and Information Science, Knowledge-Based Software Engineering: 11th Joint Conference, JCKBSE-2014, Volgograd, Russia, T. 466, pp. 585–592. 17–20 Sept 2014.

17. Husak, D.: Aspiration, execution, and controversy: reply to my critics. Criminal Law Philos. **12**, 351 (2018). https://doi.org/10.1007/s11572-017-9446-5

18. Marchetti, P., Siciliano, G., Ventoruzzo, M.: Correction to: dissenting directors. Eur. Bus. Org. Law Rev. **19**, 215 (2018). https://doi.org/10.1007/s40804-018-0101-y

Internet as a Crime Zone: Criminalistic and Criminological Aspects

Elena Prokofieva, Sergey Mazur, Elena Chervonnykh
and Ronald Zhuravlev

Abstract The emergence and rapid development of the global Internet computer network at the end of the last century have entailed an inevitable attack by terrorist organizations, criminal associations, and groups, as well as certain criminal elements, at the information and telecommunications infrastructures of industrialized countries. The chapter is devoted to one of the actual problems of modern society—the Internet as a crime zone. As a result of the research, the main measures of counteracting terrorist activity on the Internet, ways to combat hackers-fraudsters, as well as other Internet crimes are suggested. The data obtained from the study can be used to predict and counteract crimes committed on the Internet.

Keywords Internet · Internet crime · Crime on the internet · Information crime
Computer information · Information technology · Criminalistics
Forensic aspects · Criminology · Criminological aspects

1 Introduction

The Internet for any person has become a space for the free self-expression of the individual, where the role of state and law enforcement bodies, the media, and other

E. Prokofieva
Volgograd Academy of the Ministry of Internal Affairs
of Russia, Volgograd 400089, Russia
e-mail: Olenyonok83@mail.ru

S. Mazur · E. Chervonnykh (✉) · R. Zhuravlev
Academy of Management of the Ministry of Internal Affairs
of Russia, Moscow 125993, Russia
e-mail: kazanceva83@bk.ru

S. Mazur
e-mail: odir_amvd@mvd.ru

R. Zhuravlev
e-mail: odir_amvd@mvd.ru

© Springer Nature Switzerland AG 2019
A. G. Kravets (ed.), *Big Data-driven World: Legislation Issues and Control
Technologies*, Studies in Systems, Decision and Control 181,
https://doi.org/10.1007/978-3-030-01358-5_10

things is not so important. But in the modern society, there is a degradation of the Internet space; it began to change rapidly, turning from the territory of freedom of expression in the battlefield for espionage games [1].

Today the Internet, unfortunately, is a territory where you can sell tons of drugs, terabytes of child pornography, and transfer billions of flows of "dirty" money. The total absence of geographical borders and transparency, difficulties in determining the nationality of network objects, the implementation of anonymous access to its resources—all these factors make the public and personal security systems vulnerable [2]. It has become possible to commit crimes on the Internet because of the actions of individuals or groups of individuals whose purpose is to damage the information environment or its use for illegal purposes.

Internet crime is an illegal, socially dangerous act committed through the Internet [3]. Crimes on the Internet (or crimes committed with the help of this network) include: the spread of malicious viruses, the cracking of passwords, the theft of credit card numbers and other bank details, and the dissemination of unlawful information through the Internet (libel, pornographic materials, materials that incite interethnic and interreligious hostility, etc.) [4].

The research [5] explains the grading of methods of committing the Internet crimes, so when the Internet is used as a means for committing illegal actions, it is possible to speak about the commission of crimes using the Internet, and when the criminal act itself is committed via the Internet, one should speak of the actual commission of a crime through the Internet. Thus, in cases where the Internet is directly used for the commission of a crime, it is a method and means at the same time, and in the other cases, it is only the means.

2   Criminalistic Aspects in the Understanding and Study of Internet Crime

Let's turn to a criminalistic assessment of the Internet crime. In a broad sense, a criminalistic characteristic means a complex of the most characteristic criminalistic information about the features and properties of such a series of crimes that serves as the basis for the presentation of versions of the crime and the criminal personality, in our case in the Internet space, which allows us to accurately assess the situations arising in the disclosure process and investigation of computer crimes, which determines the application of appropriate methods, techniques, and tools. The Internet crimes are determined by high secrecy, a rather complex process of collecting evidence and the process of proof, a wide range of forensic signs, the absence of a unified program to combat this type of crime and a generalized judicial and investigative practice in this category of cases [6–8].

The most important from a criminalistic point of view are the following characteristics of the Internet crime:

– the way of committing a crime;

- peculiarities of trace information and situation of committing a crime;
- personal characteristics of the offender;
- peculiarities of the direct object of a criminal assault.

Criminalistic classification of crimes is actively used in the course of investigation activities, as it helps to ensure the correct understanding of the events, selection and application of criminalistic methods of investigating certain types of crimes on the Internet or the Internet crimes [6–8].

On the basis of the provisions of the Convention of European Council on Crime in the Sphere of Computer Information ETS No. 185 dated 23.11.2001 [9] the Internet crimes are divided into five groups, three of which are consistent with the chapter of the Criminal Code of the Russian Federation [10]:

- crimes against the confidentiality, availability of computer information;
- crimes committed using computer tools;
- crimes related to the creation, dissemination, possession of child pornography;
- copyright violation;
- violation of human rights on racial, religious, physical grounds, incitement to violence.

Russia has not signed the Convention [11], but among all the network crimes in our country, there is also carding and phishing.

Carding is a fraud associated with payment cards, and for the commission of this crime, there is no need to have a card. Criminals have enough details, which can be obtained by hacking the server of the online store or the personal computer from which payments were made [12].

Phishing is a fraudulent activity, because of which the confidential data of the user, logins, and passwords are in the hands of the criminal.

Considerable importance in the criminalistic assessment of the Internet crime is determining the identity, capabilities, and motives of a criminal. Criminals create and use DDoS attacks (attacks on a website whose primary purpose is to deactivate the system by filing a large number of false requests) to servers of companies, deactivate them and demand money for stopping these attacks [13]. In the Internet space there are so-called virus makers, people who create viruses inventing new and new algorithms for the intrusion of systems, bypass antivirus scanners and monitors with the aim to stand out in the virtual space, to test their own hand at writing non-standard code and non-standard software, to steal valuable documents, to get access to the system, writing destructive algorithms for mass destruction of computers [14]. There are very common fraudulent transactions of various types implemented by hackers, for example, billing for services that had not been rendered (cramming), online auctions in which the sellers themselves make bids to raise the price of the auctioned item.

Criminalistic characteristic of crimes, including the Internet crimes, is presented as a range of significant and permanent qualitative characteristics of the group of crimes identifying the main patterns of their detection and investigation. A significant contribution to the creation of criminalistic characteristics of computer crimes on the Internet has been made by V. A. Meshcheryakov in his work [15]. He gave the definition of the criminalistic characteristics of crimes committed with the use of electronic payment instruments and systems (including the Internet) determining them as a concept of the criminalistic significant and interrelated information obtained by means of special scientific research which is an essential structural element of a technique of investigation of these crimes and promotes their disclosure, investigation and prevention [16, 17].

V. A. Meshcheryakov classifies crimes in the sphere of computer information upon the following:

– the object of a criminal assault—computer information;
– the subject of a criminal assault, all crimes including the Internet as a zone of crime, are classified into crimes having a material object of assault and crimes which have no such one;
– the number of subjects of crimes committed by one person/group of persons [16].

Based on this criterion V. A. Mescheryakov identifies the following types of the Internet crimes:

– the destruction/deleting of computer information;
– illegal obtaining of computer information or violation of exclusive rights to its use;
– action/inaction aimed at the development of computer information with desired properties;
– unauthorized change of computer information.

The formation of the theory of criminalistic classification of crimes on the Internet facilitates the most correct choice of tactics and methods of investigation of any crime, including on Internet. From a practical point of view, the criminalistic classification is required to guarantee the universality and completeness of investigation and prevention of such crimes [7, 8].

In the Russian Federation, the Internet crimes are punishable by criminal, administrative and civil liability. In accordance with the current criminal legislation of the Russian Federation, the Internet crime is referred to a category of crimes in the sphere of computer information [18]. Responsibility for them is stipulated by the Chapter 28 of the Criminal Code of the Russian Federation. Online fraud, for instance, is under the Article 159 of the Criminal Code of the Russian Federation. The punishment for libel (on the Internet) is determined according to the Article 129 of the Criminal Code of the Russian Federation. More recently, these bills have been supplemented by the anti-terrorist package "Yarovoy", which regulated additional measures to counter extremism and terrorism, and toughened responsibility for crimes of this nature.

Administrative punishment occurs, for example, for propaganda or illicit advertisement of drugs and psychotropic substances in accordance with the paragraph 1 of the Article 6.13 of the Code of the Russian Federation of Administrative Offences. Violation of copyright and related rights, as well as inventive and patent rights to the information object (work) placed and distributed on the Internet, the legislation provides for civil liability in accordance with the Chapter 70 of the Civil Code, as well as for criminal liability [19].

The Internet crime is similar to the conventional categories of crimes, the only difference is that to reveal such crimes is difficult in view of the fact that the criminal and the injured party can be in different countries. The attractiveness of information technology for criminals is determined primarily by the fact that the audience of potential victims is sufficiently numerous, the territorial remoteness is obvious, criminal exposure is constant, direct contact between the potential victim and the criminal is absent, and a time gap between the beginning of active actions and consequences is provided [12, 20]. For concealment of traces of criminal activities including dissemination of illegal information, mechanisms of anonymization are used. First of all, these are anonymous centralized, decentralized and hybrid networks, while the so-called Dark (Black) Internet carries the greatest danger [21].

3 Criminological Aspects of Understanding and Studying Internet Crimes

Let's define the criminological aspect of the main indicators of the Internet crime. The most important place in the rating of the Internet crime is the analysis of its status, structure, and trends. In particular, the qualitative and quantitative characterization of criminality is the starting point of criminological research. In order to properly determine the causes, consequences, and measures of control and prevention of the Internet crime, you need to realize the scale of this type of crime. According to most authors, the statistics do not reflect the real situation of modern crime but quantitative characteristics can give an idea about the main trends in a particular state (region) for a certain period of time [22].

In accordance with the data of the Prosecutor's office of the Russian Federation, the number of the Internet crimes from 2013 to 2016 has increased by 6 times. In 2016, there were 66 thousand of the Internet crimes. In 2013, this figure was 11 thousand. Significant growth of such crimes we can see in the current year (+26%, 40 thousand). The damage from crimes committed using information and communication technologies during the first half of 2017 exceeded 18 million USD [23, 24]. Last year two-thirds of the extremist crimes and every ninth crime of a terrorist nature were committed with the use of the Internet [25].

According to the report on global risks, published following the results of the Davos Economic Forum in 2016, the risks associated with the attacks on the Internet, as in previous years, have a commensurate and very significant level [26]. Incidents of security breaches on the Internet occur in all sectors of the economy and everyday life and are inextricably linked to the manifestations of terrorism, and methods of protection and attack are the object of considerable interest of various illegal structures and groups [27].

Today Russia has not less than 2.5 thousand of the Internet sites with illegal content which spread harmful information and the Internet resources promoting extremist and terrorist activities. This fact can be explained by the low efficiency of measures to counter and prevent this type of crime. As you know, statistics do not accurately reflect the real situation in this area. The article [22] refers to some factors that affect it. The first thing cited as an example is that the Internet crime has acquired new forms of socially dangerous acts which are not covered by the articles of Chapter 28 of the Criminal Code of the Russian Federation.

For example, in the United States, the predominant Internet crime is fraud in the Internet auctions [28]. In Russia, this type of crime is also quite common, but this fact is not reflected in the official statistics. The second thing described by the researchers [22, 26] is that any system, including law enforcement, has a maximum limit of the number of crimes that it can detect and register which increases the invisibility of this type of crime. This, to some extent, determines the quality of officially registered crimes in Russia under the Chapter 28 of the Criminal Code of the Russian Federation. After all, the more limited are the possibilities of law enforcement agencies, the more often dangerous and well-organized Internet crimes remain outside of statistics. Thirdly, cases with a lot of episodes (150–200 episodes) strongly influence the statistics in this or that region. Thus, 2–3 criminal cases can result in 300–600 registered crimes [28].

4   Conclusion

Analysis of the law enforcement practice shows that technological, organizational and legal approaches are most effective for solving the problem of preventing the Internet crimes. The first one presupposes the prevention of crimes, primarily due to technical measures. The second one is related to the implementation of a variety of organizational arrangements. The third approach is based on the improvement of legal mechanisms: the improvement of the legal framework for combatting this type of crime, the optimal solution of the problems of criminalization of socially dangerous acts, the consolidation of procedural mechanisms, and so on [29, 30]. Thus, for the competent conduct of activities aimed at the prevention of the Internet crimes, it is necessary to minimize or even nullify the gaps in the criminal legislation of the Russian Federation, to evade the unrefined statistical and analytical data on a number of Internet crimes, to stop the futility of attempts to ensure the safety of computer networks solely due to protective organizational and technical measures.

Today the society has ceased to perceive the Internet crimes as innocent misconduct or fraud that is undetectable. The criminalistic and criminological evaluation of these crimes allows to correctly implement their investigation, expose and punish the perpetrators, and organize measures for the prevention of the Internet crimes involving the use of a firewall to protect the computer from hackers, to purchase and install licensed antivirus programs, to promote implementation of purchases only via safe websites, and to create strong passwords for accounts in social networks [31].

References

1. Kuru, D., Bayraktar, S.: The effect of cyber-risk insurance to social welfare. J. Fin. Crime. **24**(2) (2017)
2. Carryev, B.S., Aidarkhanov, M.B. Balafanov, E.K.: World intervision: basics of information culture. Teaching-methodical manual. Almaty **404** (2006)
3. Byers, S., Rubin, A.D., Kormann, D.: Defending against an internet-based attack on the physical world. ACM Trans. Internet Technol. **4**(3), 239–254 (2004)
4. White, R.: The four ways of eco-global criminology. Int. J. Crime, Justice Soc. Democracy **6**(1), 8–22 (2017)
5. Tarasic, N.M.: Analysis of the legal basis for combatting cybercrime. Adv. Chem. Chem. Technol. **XXX**(5), 66–68 (2016)
6. Vekhov, V.B.: Peculiarities of the organization and tactics of the crime scene search in the investigation of crimes in the field of computer information. Russ. Investigator **7** (2004)
7. Cavezza, C., McEwan, T.E.: Cyberstalking versus off-line stalking in a forensic sample. Psychol. Crime Law **20** (10) (2014)
8. Gottschalk, P.: Categories of financial crime. J. Fin. Crime **17**(4) (2010)
9. Convention on Computer Crimes. http://www.coe.int/ru/web/conventions/full-list/-/conventions/treaty/185. Accessed 06 Oct 2017
10. Criminal Code of the Russian Federation of 13.06.1996 № 63-FL, 29.07.2017. Search System "Consultant"
11. Convention on Computer Crimes. http://www.coe.int/ru/web/conventions/full-list/-/conventions/treaty/185. Accessed 06 Oct 2017
12. Manakhov, S.A.: On improving the activities to detect and investigate crimes committed using modern information technologies. In: Meeting of the Collegium of the Ministry of Internal Affairs of Russia (2014)
13. Baldwin, J.M.: Investigating the programmatic attack: a national survey of veterans treatment courts. J. Crim. law Criminol. **105**(3), 705–752 (2017)
14. Herzberg, A., Jbara, A.: Security and identification indicators for browsers against spoofing and phishing attacks. ACM Trans. Intern. Technol. **8**(4) (2008)
15. Solvak, M., Vassil, K.: Could internet voting halt declining electoral turnout? New evidence that E-voting is habit forming. Policy Intern. **10**(1), 4–21 (2018)
16. Meshcheryakov, V.A.: Basics of the methodology for investigating crimes in the field of computer information. Ph.D. Thesis. Voronezh. Voronezh State University, 387 (2001)
17. Penney, J.: Internet surveillance, regulation, and chilling effects online: a comparative case study. Internet Policy Rev. **6**(2) (2017)
18. Chen, N.-Y., Liu, Y., Zhang, Z.-J.: A forecasting system of micro-blog public opinion based on artificial neural network. J. Internet Technol. **16**(6), 999–1004 (2015)
19. Brantingham, P.J.: Crime diversity. Criminology **54**(4), 553–586 (2016)
20. Kratcoski, P.C. Edelbacher, M. (eds.): Perspectives on Elderly Crime and Victimization (2018)
21. Evdokimov, K.N.: Structure and state of computer crime in the russian federation. Legal Sci. Law Enforcement Pract. **1**(35), 86–94 (2016)
22. Dremlyuga, R.I.: Internet Crimes: Monograph. Publishing House of Far East University, Vladivostok, pp. 240 (2008)
23. Cabaye, M.: Studies and Surveys of Computer Crime. Northfield **2** (2001)
24. Ifrah, L.: Cybercrime: Current Threats and Trends. COE (2008)
25. Klinger, D.: Spreading diffusion in criminology. Criminol. Public Policy **2**(3), 461 (2003)
26. Global Risks Report, 2016. The World Economic Forum. The Global Risks Report 2016.pdf http://book.itep.ru/depository/annuals/. Accessed 06 Oct 2017
27. De Sanctis, F.M.: Football, gambling, and money laundering. A Global Crim. Justice Perspect. **I**, 213 (2014)
28. Gercke, M.: Understanding Cybercrime: A Guide for Developing Countries. ITU (2009)
29. Nomokonov, V.A., Tropina, T.L.: Cybercrime as the new criminal threat. Criminol.: Yesterday, Today, Tomorrow **24**(1), 45–55 (2012)

30. Evdokimov, K.N.: Structure and state of computer crime in the Russian federation. Legal Sci. Law Enforcement Pract. **1**(35), 86–94 (2016)
31. Quyên, L.X., Kravets, A.G.: Development of a protocol to ensure the safety of user data in social networks, based on the Backes method. In: Communications in Computer and Information Science, 466 CCIS, pp. 393–399 (2014)

Implementation of the Law Enforcement Function of the State in the Field of Countering Crimes Committed Using the Internet

Vyacheslav Urban, Viktor Kniazhev, Anatoly Maydykov and Elena Yemelyanova

Abstract Due to the wide spread of information and communication technologies, the nature of the crime is changing. At present more and more crimes are committed using computer technology and the Internet. The purpose of this chapter is the analysis of crimes committed using the Internet and the elaboration of a system of measures to combat such crimes. The main directions for increasing the effectiveness of countering crimes committed using the Internet, including those related to the development of international cooperation in the criminal process, are proposed. The data obtained can be used for prevention and effective combatting crimes committed using the Internet.

Keywords Internet · Internet crime · Crime on the internet · Information crime
Computer information · Information technology

1 Introduction

One of the key functions of a state is law enforcement that is designed to ensure the survival, preservation, and development of society as an integral system. The basis of the law enforcement function is the activity on the protection of rights and freedoms

V. Urban · V. Kniazhev (✉) · A. Maydykov · E. Yemelyanova
Academy of Management of the Ministry of Internal Affairs of
Russia Moscow, Moscow 125993, Russia
e-mail: 1port@mail.ru

V. Urban
e-mail: policaj@mail.ru

A. Maydykov
e-mail: odir_amvd@mvd.ru

E. Yemelyanova
e-mail: katolean@gmail.com

© Springer Nature Switzerland AG 2019
A. G. Kravets (ed.), *Big Data-driven World: Legislation Issues and Control Technologies*, Studies in Systems, Decision and Control 181,
https://doi.org/10.1007/978-3-030-01358-5_11

of individuals, organizations, and as a whole of a society and a state. The main form of implementation of the considered function is law enforcement activities that include such directions as ensuring public safety, protection of public order, activities aimed at detection and investigation of crimes, administration of justice, criminal and penal activities, etc.

In the framework of this chapter, we shall limit the scope of the study by criminal procedure activities of the law enforcement bodies in terms of organization of counteraction to crimes committed with the use of the Internet.

The number of the Internet users in Russia is constantly growing. Thus only in the Russian segment of the Internet in 2016, there were more than 80 million persons, while the turnover of goods and services using the Internet in 2015 was equivalent to 2.3% of the gross domestic product, and it currently keeps a tendency to growth [1]. The Internet resources are very diverse—electronic media, websites of executive bodies, sites of submission of online ads, social networks, etc. Accordingly, a person has the opportunity to carry out virtually the same actions via the resources available on the Internet as in real life. At the same time, the virtual space is under much weaker control from the state, rather than the real environment.

Due to the wide spread of information and communication technologies, the nature of the crime is changing. At present more and more crimes are committed using computer equipment and the Internet.

2 Internet Crimes Features and Classification

Internet crime is growing at the fastest pace all over the planet. Information infrastructure allows criminals to act covertly, not leaving their coordinates; to commit criminal acts from the territory of any state (where there are technical possibilities), and to work with subcontractors at an international level.

The following features of the Internet contribute to its active use for criminal purposes:

(1) potentially unlimited audience for the dissemination of information;
(2) problems with the organization of effective state control. Because of the nature of the organization of information-telecommunication networks, particularly the Internet, their cross-border and international character, the state has limited capacity to regulate and control the use of the network resources;
(3) high anonymity, protection, secrecy of communication and information exchange in the network;
(4) free of charge or low cost of developing (receiving) and using network resources and communications means;
(5) dynamism of the network technologies that allows members of the criminal organizations to create information resources and use network communications faster than the law enforcement bodies have time to adequately react to this.

The UK experts distinguish similar directions of the influence of modern information and communication technologies on the growth of crimes committed on the Internet [2].

According to estimates made by the "Symantec" company, the total revenue of cybercriminals is more than 110 billion dollars a year. In this case, the collateral damage resulting from the commission of these crimes (the time spent for recovery, downtime costs, etc.) can be estimated at an additional $ 274 billion. Accordingly, the total damage from cybercrime on a global scale is about $ 388 billion per year. At the same time, the global world turnover of marijuana, cocaine, and heroin is about $ 288 billion, which is 26% lower. Other organizations, regarding the world economy's losses from cybercrime, give even more impressive figures (445 billion dollars [12], 500 billion dollars [19]). According to Microsoft, by 2020 the world economy will lose 3 trillion cumulative losses from computer crime. dollars [3]. By the estimation of the Microsoft Corporation, the cumulative losses of the global economy from the computer crime will have been about 3 trillion dollars by 2020 [3, 10].

According to the European Council Convention on Cybercrime, signed in Budapest on 23 November 2001, the concept of cybercrime covers the following crimes [11]:

(1) crimes against confidentiality, integrity, and availability of computer data and systems, including illegal access to computer systems, illegal interception, data interference, interference in systems and abuse of equipment;
(2) crimes associated with computers, including forgery and fraud committed using computers;
(3) crimes associated with content (particularly child pornography);
(4) crimes associated with infringement of copyright and related rights.

In turn, the modern Russian criminal legislation provides sanctions only for the acts related to the violation of security of computer information, that is for crimes in which the objects are the public relations arising in the sphere of the computer information or ensuring the safety of computer information [4].

At the same time, computer-related offenses are currently committed not only with the use of information and communication technologies but also via the modern types of computer programs or software and hardware that can be used as means or instruments for their commission.

In the fair opinion of I. M. Rassolov, now computer-related crime is a criminal industry in which there are scammers, hackers, extortionists, pedophiles, pimps, human and drug trafficking, as well as many other offenses.

According to the definition of the Federal Criminal Police Department of Germany, the concept of cybercrime covers all crimes committed with the use of equipment for collecting, processing and transmitting information or communications, or crimes committed against them [5].

Accordingly, cybercrime can be conditionally divided into two main groups: crimes directed against computer hardware or the Internet network, and crimes committed using a computer or the Internet [6].

So, there are crimes committed with the use of information and communication technologies in the economic sphere. These include fraudulent acts of persons associated with the hacking of users' profiles in social networks and subsequent distribution of messages to "friends" of the victim of the burglary asking for the transfer of funds under various pretexts. Fictitious sale of goods (services) can be added to this group.

The main instruments of such crimes, as illegal turnover of means of payments, illegal banking activities are also information and communication technologies. Thus, committing these crimes involves conducting banking transactions related to the transfer of funds. In this case, payment orders are sent mainly via the Internet that greatly accelerates and facilitates criminal activities and creates certain obstacles for documenting this activity.

The number of cases of crimes against sexual freedom and sexual inviolability, in particular against minors, constantly increases.

Thus, in the Russian Federation, there is yet one enforceable guilty verdict for committing sexual acts (Article 132 of the Criminal Code of the Russian Federation) using the Internet with respect to the victim who has not reached the age of fourteen [7]. A man, through a social network VKontakte, having met a girl under 14 years of age, began to show photos of nude male genital organs, accompanying them with text messages with suggestions for sexual acts for monetary compensation. In addition, there are websites in the network that contain pornographic materials involving minors.

3 The Nature of the Unlawful Information on the Internet

3.1 "The Death Groups"

Over the past few years, "the death groups" have been actively developing on the Internet, designed to induce children and adolescents to commit suicide. In order to join the group, it is necessary to perform certain tasks, mainly related to photographing oneself in dangerous places, for example, on the edge of roofs of high-rise buildings, on railway rails, etc. In addition, the members of the group are taught that they are of no use to anyone, that there is a happier world, where one should strive to end up. The psychological impact on members of groups consists, basically, in the exploitation of their need to belong to a significant secret group, to feel their uniqueness. The content of these groups contains images (photos and videos) of methods of suicide, the killing of animals, with this video broadcast accompanied by anxious, depressive music with background cries, crying children. With regard to the adolescents, moral and family values are manipulated, feelings of guilt are instilled, and threats of reprisal against their relatives are often voiced. Because of the fear of becoming the cause of death of loved ones, as well as under the psychological influence, a teenager prefers to commit a suicide.

Currently "groups of death" began to actively develop their activities in the European countries. The official website of the Ministry of Internal Affairs of France is dedicated to the Internet threats; information is posted on the dangers of "suicidal games from Russia", functioning in social networks [8].

3.2 Extremist and Terrorist Content

Members of various destructive organizations actively use the Internet for the dissemination of materials of extremist nature, for the recruitment of new members, and for the coordination of illegal activities. In addition, the Internet is actively used as a means of communication, remote management and training of terrorist activities, and finding additional sources of funding.

According to various reports, currently, in the world, there are about 5000 Internet sites containing extremist (terrorist) content. As for the Russian sites of an extremist orientation, more than 90% of them are outside the Russian legal jurisdiction.

To achieve their goals the representatives of extremist and terrorist organizations use the possibilities of social networks (VKontakte, Facebook, Twitter, Odnoklassniki) where active propaganda work is conducted. Currently, the number of crimes committed using the Internet prevails and has a steady upward trend. In the structure of extremist crime, the proportion of such acts is virtually two-thirds [8].

3.3 Illegal Drug Trafficking and Gambling Establishments

Communication possibilities of the Internet are also actively used to commit crimes related to illegal drug trafficking. Today the contactless sale of drugs is most widespread. Thus, the contact of a consumer with a distributor takes place by correspondence or calls via social networks and instant messaging applications (e.g. Skype, Viber, WhatsApp, etc.); then the payment is made through various payment systems. After payment is made the distributors inform the consumers about the hiding places with the drug substance. The substance is usually placed in such a way so that to minimize the possibility of its detection by unauthorized persons [14].

In addition, the law prohibits implementation of activities related to the organization and conduct of gambling using information and telecommunication networks (including the Internet). In the territory of the Russian Federation gambling establishments can operate only in specially designated gambling zones. At the same time, many entrepreneurs register and carry out the organization of gambling from the territory of foreign states, bypassing the Russian legislation. To prove the fact of organizing illegal gambling activities, it is necessary to have access to the appropriate servers, which, as noted above, are mainly located in the territory of foreign states. In this regard, to prove the fact of criminal activity is quite problematic.

4   Solving the Problem of Internet Crimes Counteraction

Thus, at the present time, there is a clear tendency to increase the number of crimes committed using the Internet (information and communication technologies); and in the near future, this trend is not expected to change.

The specific characteristics of these crimes are determined by the use of high technologies, the necessity of possessing a certain level of special knowledge and special tools to commit them. These circumstances contribute to high latency of these crimes that significantly complicates their identification and documentation, as well as the organization of countering them.

These methods of counteraction can be conditionally divided into internal and interstate ones. Internal methods of counteraction include improvement of forensic tactics and investigation techniques, as well as the development of new requirements for the collection, verification, and evaluation of evidence, in the investigation of crimes committed using the Internet.

With regard to the Russian Federation, it seems advisable to pay considerable attention to the development of tactics for carrying out such investigative actions as the inspection of the scene of an incident, the search (seizure) of telephone and other negotiations, interrogation.

The specifics of the inspection of the crime scene, as well as the search or seizure, requires thorough preliminary training, the involvement of experts of the appropriate profile, the use of special technical devices and software (in addition to standard technical and forensic tools), the adoption of measures aimed at ensuring the preservation of computer information.

It is also necessary to develop recommendations on the seizure, packaging, and transportation of computer equipment and other storage media found during these investigative actions.

Conducting control over the telephone and other negotiations, for cases of this category, can help identify locations of used computer equipment, other illegal items, documents (in particular, electronic ones).

During interrogation, considerable attention should be paid to the use of computer-based communication technologies for criminal purposes.

Besides, improving tactics and methods of investigation it is expedient to adopt additional measures aimed at improving the safety of the Internet space. So, it is necessary to study the issue of mandatory identification of the Internet users. Similar proposals are being actively discussed in other countries, particularly, in the USA, where in April 2009 a bill on cybersecurity developed by the National US intelligence was registered [9].

Given that the principle of territoriality is practically inapplicable to the Internet, the fight against such crimes only at the national level is ineffective. Due to the fact that in many cases the providers of computer and information and telecommunication services are registered in the territory of foreign countries, one of the most important issues in the field of countering crimes committed using the Internet is the need to improve international cooperation, including in the criminal process.

5 Conclusion

Currently, the main form of the international cooperation in the criminal procedure area is the sending of requests for legal assistance. The procedure for interaction between the Russian courts, prosecutors, investigators and inquiry agencies with the relevant competent authorities and officials of foreign countries is regulated by Chaps. 53, 54 and 55 of the Code of Criminal Procedure of the Russian Federation.

International legal assistance in criminal cases may be provided when the need of executing such investigative actions as inspection, search (seizure), interrogation, examination or other procedural actions arise in the territory of a foreign state. In particular, when an accused (suspect) hides in the territory of another state, or in cases, when witnesses, victims or other participants of criminal proceedings, as well as documents necessary for the investigation, are in the territory of another state.

It is necessary to develop international cooperation in criminal proceedings especially for crimes committed using the Internet. So, it is advisable to provide for a simplified procedure for cross-border cooperation with suppliers of computer and information and telecommunications services with the aim to obtain data about subscribers, traffic, and content of communications. The main purpose of the development of simplified orders should be the speed of execution of requests for legal assistance. That means that the requested information should be provided in the shortest possible time that directly affects the efficiency of counteraction to the above criminal acts.

In addition to interaction in the criminal processing area, it is necessary to develop cooperation at the level of global technology corporations. An example of such cooperation may be the agreement reached between YouTube, Microsoft, Facebook, Twitter, and Google in October 2017 on consolidating efforts in the fight against terrorist content in the network [13].

Thus, for the effective implementation of the law enforcement functions of a state in the sphere of counteraction to the crimes committed with the use of the Internet it is necessary to improve criminalistics tactics and methods of investigation, to develop new requirements for collecting, checking and evaluating evidence for the cases of this category, as well as further development of international cooperation including in the sphere of criminal process.

References

1. Decree of the President of the Russian Federation of 09.05.2017 No. 203 "On the Strategy of Development of Information Society in the Russian Federation for 2017–2030". "Code of Laws of the Russian Federation", 15.05.2017, No. 20, Article 2901
2. Von Behr, I.: Radicalisation in the Digital Era. https://www.rand.org/content/dam/rand/pubs/r esearch_reports/RR400/RR453/RAND_RR453.pdf (2013). Accessed 1 Jul 2018
3. Annual damage from cyber-attacks is $400 billion. anti-malware https://www.anti-malware.r u/news/2016-10-13/21196#. Accessed 1 Jul 2018

4. Rassolov, I.M.: Cybercrime: concept, main features, forms of manifestation. Legal world. 2 (2008)
5. Zygmunt, O.A. Petrovskiy, A.V.: Bundeskriminalamt. Cybercrime: Bundeslagebild 2011 Wiesbaden, P.5. Quote of: Cyber and Internet Crime in Germany and Russia: possibilities of comparative research. Leg. Sci. Law Enforc. Pract. **4**(34), 180–188 (2015)
6. Marbeth-Kubicki, A.: Computer- und Internetstrafrecht. Munchen: Verlag C. H. Beck. Quote of: Zygmunt O. A. Petrovskiy A. V. Cyber and Internet Crime in Germany and Russia: possibilities of comparative research. Leg. Sci. Law Enforc. Pract. **4**(34), 180–188 (2015)
7. Popova, L.V., Mikhailov, V.V.: Criminal liability for committing acts of a sexual nature using a network. Zakonnost **2**, 44 (2016)
8. Official website of the Ministry of Internal Affairs of France. https://www.internetsignalemen t.gouv.fr/PortailWeb/planets/Actualites.action;jsessionid (2018). Accessed 1 Jul 2018
9. Official website of the Congress of the United States of America. S. 773: Cybersecurity act of 2009. [Electronic resource] URL: https://www.congress.gov/bill/111th-congress/senate-bil l/773
10. Pokar, F.: New challenges for international rules against cybercrime. Eur. J. Crim. Policy Res. **10**(1), 27–37 (2004)
11. Oxford English Dictionary. Oxford University Press. http://www.oxforddictionaries.com/defi nition/english/cyber (2015). Accessed 5 May 2015
12. The Global Risk Report 2016. World Economic Forum. http://book.itep.ru/depository/annual s/TheGlobalRisksReport2016.pdf (2018). Accessed 1 Jul 2018
13. Facebook, Microsoft, Twitter, and YouTube launch anti-terrorism partnerships. https://www. theverge.com/2017/6/26/15875102/facebook-microsoft-twitter-youtube-global-internet-foru m-counter-terrorism (2018). Accessed 1 Jul 2018
14. Torgovchenkov, V.I., Ivanov, S.A.: Features of the Prevention of Non-Contact Ways of Drugs Sales in the Russian Federation. Laws of Russia: Experience, Analysis, Practice **12**, 85 (2016)

Counteraction to E-Commerce Crimes Committed with the Use of Online Stores

Olga Dronova, Boris P. Smagorinskiy and Vladislav Yastrebov

Abstract Expanding the opportunities for introducing information and telecommunication systems in all areas of modern society, along with a positive effect, inevitably leads to a number of negative consequences associated with the misuse of innovative aids for the purpose of committing crimes. The emergence of new goods in the consumer market, in conjunction with the impersonal information aids for the implementation of their turnover, often exceeding the limits of national economies, require the development of effective proactive preventive measures aimed primarily at the informational support of the virtual Internet market. This support can be based on an electronic platform containing blocks of legal information on the regulation of this segment of trade turnover, lists of names of bona fide Internet shops and those whose activities are blocked by court decision. This platform also allows systematizing information about the technical and legal properties of the tested goods, their compliance with established requirements. Information obtained from various sources (public services, non-profit human rights organizations and consumers) on qualitative and quantitative characteristics allows one to obtain systems of properties of cognizable objects, information about which is a means of solving many searches and cognitive and verification tasks and is also an independent source of information.

Keywords Internet of things · Properties of goods · Counterfeit
Consumer market · Information system

O. Dronova (✉) · B. P. Smagorinskiy
Volgograd Academy of the Ministry of Internal Affairs
of Russia, 400089 Volgograd, Russia
e-mail: nio-va@rambler.ru

V. Yastrebov
Academy of Management of the Ministry of Internal Affairs
of Russia, Lomonosov Moscow State University, 125993 Moscow, Russia
e-mail: 18031937@rambler.ru

© Springer Nature Switzerland AG 2019
A. G. Kravets (ed.), *Big Data-driven World: Legislation Issues and Control
Technologies*, Studies in Systems, Decision and Control 181,
https://doi.org/10.1007/978-3-030-01358-5_12

1 Introduction

Currently, Russia has actively advanced in many areas of the digital economy, placing it as the basis of public administration, the economy, business, the social sphere and society as a whole. The concept of the information society provides for the dominance of new technologies, including blockchain in eight areas affecting state regulation, information infrastructure, research and development, staff and education, information security, public administration, smart city, digital healthcare. According to the results of 2016, the Russian market of commercial data storage and processing centers amounted to more than 14.5 billion rubles.

In the countries of the European Union this approach to the organization of the economy has been applied since 2000 in different periods, the terms "Internet economy", "digital economy", later "electronic (networked) economy" were applied to it. Based on data provided by the European Information Technology Observatory (EITO), revenues from the use of information and communication technologies (ICT) in 2017 have increased by 1.8% compared to 2016 and amounted to 683 billion euros. A significant increase in the sales of IT equipment, IT services and software provides grounds for talking about a significant digitization of the economies of the countries of the European Union [1]. The technological sector of the "Internet of Things" (IoT) deserves a special consideration as it allows to combine smartphones, laptops, computers and other equipment with the cloud platform to which they are connected. They are inextricably linked to the systems, fixing protocols of the interaction of sensors, carrying out their storage, processing, and protection of information received. Objects communicate via wireless communication using Wi-Fi, Bluetooth, LoRaWAN, BLE, Ethernet, RFID, ZigBee etc.

The practical potential of using the "Internet of Things" gives broad opportunities to all participants of the consumer market (producers of goods and services, transport companies, sellers, and consumers) to receive information about all stages of the consumer chain. Each of its directions provides for the possibility of fixing a multitude of parameters: remote monitoring of soil condition and livestock health, serviceability and workload of production equipment, qualitative and quantitative parameters of the finished product, transport and retail logistics, monitoring of the sales cycle, control of the authenticity of products participating in the commodity turnover, crowdsourcing for tracking customer feedback [2] and fixing consumer preferences [3], etc.

According to the estimates of telecommunication equipment manufacturers by 2018, the sensors and devices used in the "Internet of Things" (IoT) will become the largest category of connected devices. The Russian market of the "Internet Things" is also actively developing. According to the "Direct INFO", the total size of the Russian market of the "Internet Things" amounted to 17.9 million devices in 2016 and grew by 42% compared to 2015. By 2021 the total number of IoT devices will grow to 79.5 million. The total potential of the Russian market is estimated at 0.5 billion devices [4].

2 The Main Elements of Criminalistics Characteristic

2.1 The Circumstances of Committing

Considering the external and internal factors characteristic both, for the activities of the enterprise and for the strategic needs of the market, one can get a holistic view of the consumer product taking into account innovations [5]. Examples of new types of goods include the creation of smart clothes capable of revealing the initial stages of certain diseases; shoes that are themselves laced up, measuring body temperature, fixing the distance traveled; household appliances reading information labels on goods that allow them to identify and determine qualitative characteristics etc. There are also technological products of the higher level, such as technologies of unmanned vehicles, drones-quadcopters, smart home systems, which allow to automate the maintenance of objects [6] (lighting, video surveillance, heat supply, etc.) and other technological novelties.

At the same time, insufficient attention is paid to the security of this group of products from outside interference and unauthorized hacking. Such potential sectors of the digital economy as "smart cities", "virtual assistants", "Internet of things", "robot surgeons" and other technological products are considered as potential cyber threats. There are threats of large-scale DDOS attacks on various web resources able to manage the configuration and services of devices and to control the launch of the service and the programming of operations [7].

However, the widespread introduction of innovative consumer products in everyday activities has not yet received its mass distribution. The use of information technologies for the purpose of controlling qualitative and quantitative trade turnover now mainly refers to the desire of the producer to improve the economic performance of the enterprise, and in some cases is envisaged in order to comply with the fiscal interests of the state.

The essence of an innovative product is often perceived by the user as a specific subject with a clear form and structure, but in the modern technological society this concept goes beyond the simple invention [8] and extends to the ways of its transfer to a wide range of consumers. The global trade and social information networks, acting as impersonal informational means of carrying out trade turnover that go beyond national economies, are providing the basis for combining a significant amount of information about a product and its provision to the consumer market.

According to sociological research in Russia, 84 million people are Internet users, with a steady increase in the number of people using smartphones (42.1% of the country's population) and laptops (19%). The volume of the Internet market for 2017 already exceeds 1 trillion, while there is an increase in non-cash transactions made through the services of Samsung Pay, Apple Pay and Android Pay using the module of contactless data transmission (NFC). According to the estimates of the Association of Internet Trade Companies, the virtual turnover of goods and services accounts for 36% of the volume of the Russian digital economy [9].

This segment of the economy, which in some cases is known as the global sphere of "e-commerce" [10], is called to interrelate state, municipal structures, organizations and individuals with regard to production, exchange and consumption of goods and services carried out on a global scale, in whole or in part in the virtual space and aimed at obtaining personal or property benefits from the socially useful activities within the framework of the Russian and international legislation [11].

A significant growth in the development of this segment of the economy, coupled with the continuous updating of information technologies, has inevitably led to the emergence of new types of unlawful acts, including in the consumer market, carried out both by unscrupulous sellers, intermediaries, and buyers, either intentionally or by virtue of low legal literacy of acquiring goods that belong to the categories that violate the established requirements (counterfeit, falsified, non-compliant with the safety requirements, products in violation of the requirements of national standards, technical conditions, etc.). These actions often cause substantial harm to the life and health of consumers, damage fair competition in the field of entrepreneurship, and negatively affect the economic, political and information spheres of the country's activities.

At the same time, many aspects of this activity remain outside the state legal regulation, due to the backlog of the legislation from the actual development of information technologies. This fact result in the disability of law enforcement agencies to effectively prevent and counter new types of illegal activities, which are further complicated by inadequate training of the staff of the relevant departments in matters of electronic commerce, control of the Internet space, identification of network users, operations with the integrated, dedicated and cloud options, etc.

Meanwhile, the violations in the consumer market in the Internet trade segment are very high and according to experts' estimates fall on 1 of 4 purchases, while the problem is complicated by the time intervals between the transaction and the time of the detection of the illegal activities which can result in the full deletion of the transaction traces.

2.2   Classification of Victims

Illegal actions in the sphere of virtual goods turnover can cause damage.

To the consumer (an individual) who does not receive the goods after payment of its value, or purchases products completely or partially differing in their consumer and value characteristics from the declared in the Internet stores, which cannot be exchanged, carry out warranty repairs, return to the seller for the reasons of the lack of data about the real producer of the goods. Additional problems are created by the lack of the possibility of the feedback from the sender (which can be both, the seller (intermediary) and the manufacturer himself), as well as by the procedural difficulties of the actual initiation of civil lawsuits against foreign companies, domain owners. These actions can additionally be accompanied by the write-off customer's funds in excess of those provided for in the order and paid by means of the payment cards and electronic wallets used during the transaction;

To the manufacturer (enterprises of all forms of ownership), which is damaged in the sale of copies of goods using someone else's trademark, service mark, the appellation of origin. In some cases, unscrupulous "entrepreneurs" create clones of the original sites of manufacturers of well-known brands (single-page websites) that differ by several signs of the domain name and offer "similar" products with a different value and internal properties. These actions of unscrupulous entrepreneurs require legal producers to develop strategies and allocate significant financial resources to protect the incomes and intangible assets from the actions of the main strategic groups of manufacturers of counterfeit products (imitators, scammers, smugglers etc.), each of which violates current legislation by changing a certain set of the initial properties of products [12].

To the state, due to the insufficient level of control over the importation into the country products (under the guise of consumer goods) whose turnover is not regulated by law, but is capable of harming the life and health of consumers (an indefinite group of persons) or is prohibited from free circulation. Besides, the importation of a significant amount of goods of low cost and low quality impedes the development of the domestic producer, and the budget receives less than the corresponding duties and tax deductions from the distribution of goods by foreign companies. Illegal actions adversely affect the formation of the public opinion on the possibility of unpunished committing of economic offences in order to obtain benefits, thereby contributing to a decrease in the level of trust in the institutions of state power and the law in force, as well as violating consumers' rights to be provided with the high-quality products [13].

These offences become possible due to the unresolved number of legislative issues, the lack of practice of law enforcement agencies in identifying and suppressing crimes in the virtual consumer market, as well as the insufficient level of legal literacy of consumers and the difficulties in initiating criminal cases against unidentified defendants (even if data on them are placed on the sites which sell counterfeit products), creating a threat to an undefined circle of persons.

The problematic issues are:

- the uncertain economic and legal status of virtual information;
- the absence of a legally fixed concept of commercial activities carried out using the information and telecommunications network;
- the unregulated responsibility for the legality and reliability of advertising and other information provided by the owners of online stores, as well as by the social network operators;
- the lack of requirements for individualization and identification of the organizations (especially foreign ones) which carry out activities using the Internet space, as well as the sale of goods for various purposes without complying with the procedure for their certification;
- procedural difficulties with the implementation of measures to protect various types of intellectual property;

– the lack of established practices to protect the rights of producers and consumers of products and services sold remotely.

The solution of these problems requires a systematic approach to the study of the essence of e-commerce itself, the systematization of the directions of the illegal activities carried out in the sphere in question, the bringing of blanket legislation regulating the sphere of trade to a single legal field for the use of information technologies.

2.3 The Subject of a Criminal Encroachment

The study of crime in the consumer market, traditionally considered as an integral element of economic crimes, makes it possible to unite the totality of acts in the performance of which the object is social relations in the sphere of any kind of turnover, and the observance of a set of concomitant requirements ensuring economic independence and state security, observance of constitutional rights and freedoms of citizens, life and health of the nation. In this context, consumer goods, raw materials, technological characteristics and means of production, individual means of protection, legal and technical properties inherent in the commodity throughout its life cycle, logistic schemes, ways and means of delivering it to the consumer can be considered not only as a matter of criminal assault, but also as means and instruments for the commission of a crime.

At the same time, the illegal activities in the sphere of the consumer market under consideration are committed in the virtual space, regardless of the location of goods in the legal or illegal trade turnover, using computer information blocks, means of their creation, storage, processing and transmission (computers, smartphones, cash registers, ATMs, payment terminals and other computer devices), information and telecommunication networks, allow us to talk about referring them to the segment of cybercrime, and the fact that they have a negative impact on social relations in the sphere of information technologies and the safe operation of computer information [14]. Infringements in the field of IT security in the area of virtual goods turnover are most often encountered in the field of the loss of data on payments made; the illegal distribution of intellectual property, mainly represented in the form of substances, not having a specific material form and volume expression, which requires the adaptation of them with the use of additional software; the loss of information of client databases and information about their personal data.

However, consumer goods, as well as operations and services sold remotely, will continue to be considered as the main objects of criminal assault, by presenting them in online or offline environments. At the same time, trade is often carried out both in the sphere of selling legal products and in the field possessing the status of limited or prohibited turnover.

Currently, the Russian Federation provides for three modes of circulation of goods, which are differentiated into freely alienated, limited in circulation and withdrawn from circulation. The first category includes a wide range of consumer goods for food and household purposes, corresponding to the legal and technical requirements for their production and circulation in the market. However, there are some restrictions for the ways of selling some categories of goods; in particular, it is prohibited to sell medicines, alcohol products, technical aids intended for receiving information in a covert way, etc. through Internet stores. There is also a ban on the remote sale of goods made of the precious metals and stones, which in practice is violated by many online stores, in some cases imitating not the direct selling of goods, but their reservation with subsequent redemption.

The second category of goods for the legal presence in circulation provides, in addition to the above requirements, the observance of a number of conditions. These include the need to license certain types of activities; state control over the distribution of quotas for the export and import of a number of goods; conformity of subjects of turnover of certain goods to existing requirements; compliance with the procedure for passing state registration and entering into the relevant state registries, on the basis of laboratory tests of the products, goods and materials which are first produced or imported into Russia.

In addition, the state can introduce measures of restrictive economic and administrative nature, designed to facilitate import substitution and development of the domestic market.

Goods withdrawn from the circulation cannot be items of free circulation. They may include products that violate the technical and legal properties of the original goods and those to which the regime demands are applied, conditioned by the interests of the state and the threat to public safety.

Taking into account the peculiarity of the method of implementation and the status of being in circulation, all the articles of criminal infringement can be divided into those possessing physical form and the ones presented in the form of digital multifunctional products used by means of software.

Regardless of the form of expression and methods of sale, all consumer goods entering the commodity turnover in the territory of the Russian Federation must be assigned to a specific kind, class, subclass, groups, subgroups, types, subspecies, categories, subcategories, taking into account the main purpose of the product or the main group of consumers (medical, cosmetic, children's, diabetic, etc.). At the same time, they must possess certain qualitative and quantitative characteristics, technical and technological properties that must comply with the established requirements of the national standards, technical regulations, etc. So, the commodity turnover is possible only if the state's fiscal interests are observed.

However, the compliance with these requirements with the broad technological capabilities of methods of obtaining virtually any product, especially from the foreign trade sites, is almost impossible. To date, there are no requirements for the registration with the tax authority for a domain name used by the seller for the sale of goods, works or services, as well as for legal entities, and individual entrepreneurs engaged in online trading using e-mail of their stores. Law enforcement agencies face significant difficulties in identifying and suppressing illegal activities of individuals using social networks (Instagram, Facebook, etc.) and auction sites for the distribution of counterfeit goods: branded bags, luxury clothing, which in some cases are presented in the form of things previously used, which significantly complicates their identification by the manufacturer in online mode.

A consumer who realizes that he is acquiring a knowingly counterfeit product or banned for free circulation products will receive them with a probability of more than 90%, except for cases of fixing information about them by the federal customs service and stopping their importation into the country. For consumers who plan to purchase original benign goods can partially be protected from the acquisition of counterfeits by the information blocks of reviews placed on the page of the selected products that contain information on qualitative and quantitative characteristics, their compliance with the declared samples, delivery times and types of payment. However, this information is not enough, and in some cases, it is unreliable, corrected or completely formed by the owner of the online store.

3 Solving the Problem of Information Obtaining

As an active means of counteracting the illegal sale of consumer goods using the Internet space is an electronic platform [15] where there is information on the legal nature regulating the sphere of virtual turnover, data of the main domain names of the official online stores, as well as the information on those sites which are blocked by a court decision and those improper sellers whose names are included in the list of the Russian Federal Supervisory Service in the Sphere of Communication, Information Technologies and Mass Communication as banned. This information and reference system is intended for all market participants (producers, sellers, consumers) [16] and law enforcement agencies. In addition to the above directions, it provides for the integration of information on the characteristics of goods that are sold by the interested producers or exporters with data on identified falsified goods available in the law enforcement agencies due to the investigation of criminal or administrative cases.

The implementation of the creation, deployment, and functioning of the Internet information and reference system should be undertaken by the structures within the framework of state and public protection of the rights of consumers and entrepreneurs jointly with producers, representatives of rightsholders and law enforcement agencies.

The content of the information and reference system stipulates the unification of a significant amount of information in a common resource on a single hardware-software platform that must be unified and typed for the entire array. To solve a large volume of tasks, it is necessary to use an architecture with distributed computing resources and integrated banks and databases, which will allow to concentrate information in one storage, provide the ability to convert it when changing software shells and interfaces, and ensure the distributed simultaneous access to this information.

The process of forming arrays can be represented in the form of the following algorithm:

1. Study of information on goods most in demand in the consumer market that do not meet the established requirements, which are accumulated in departmental statistical data of various divisions, consumer appeals to the territorial services for supervision of consumer rights protection and human well-being, as well as open surveys of non-profit organizations, such as "The Russian System of Quality", "The Russian System of Control" etc. At the same time, the most frequently encountered items of violations are commonly highlighted, for example, household appliances, communication equipment, light industry products, alcohol products, software, pharmaceutical and cosmetic products.
   In this case, the activities of state bodies and non-profit organizations are aimed at preventing this type of offense, since consumers can use the array of data before purchasing the necessary goods. On the other hand, producers, seeing that their goods are often counterfeited by unscrupulous "competitors", will be able to develop and implement new systems for identifying their products.
2. The results of the summarized information shall be sent to the right holder or producer of the product with the proposal to disseminate this product protection information containing descriptive and illustrative material in accordance with the established unified list including labels, holograms, markings, package requirements for the final product, etc.
3. Moderator of the information-reference system should place the above information in an array on a separate page in the information folder. All information can be systematized in separate electronic pages and be presented to the consumer in several versions of the data aggregation: by the name of the manufacturer (trademark) or by the type of goods (means of communication, household appliances, branded things, etc.). It is advisable to place an illustrative image of the product from different angles in the product information page, highlighting the features of the protective elements, accompanied by the data on the manufacturer, the country of manufacture, the time of commencement (removal from) the production of products, the individual features of the product (depending on the type, it can be volume, size, color scales, the number of items in a set, etc.). It is required to include the information on the additional technical means of product protection (presence of codes under the scratch layer, use of QRGL, RFID technologies, etc.) which must be reflected with a separate reference to the rules for self-identification of goods by these systems.
4. Law enforcement agencies should use the information of this data array in preparation for the implementation of the activities aimed at identifying and fixing the dissemination of illegitimate products, including those sold by the Internet shops.

In addition, the information on the most frequently encountered ways of falsifying the main protective elements of products, as well as the variants of goods-simulators of products of a particular manufacturer and data of the Internet store where they are presented, can be placed in the individual product page.

5. As a separate possibility of replenishing the resource, it is possible to provide the consumer with the opportunity to post reviews about the purchased product without the right to make changes to the main page. Such feedback will help to

obtain additional information from the virtual trading platforms selling goods of inadequate quality.

4 Conclusion

A comprehensive understanding of the direction of the implementation of counteracting inappropriate consumer goods sale through the use of online and offline trading is possible after a detailed study of the main stages of the manifestations of illegitimate goods in circulation. Modern technological capabilities make it possible to trace the main stages of the production of goods and to effectively protect the supply chain from the penetration of counterfeit goods and copyright piracy [17]. These methods additionally include the need to monitor compliance with the terms of the contract for the production and sale of goods, the identification of suppliers and customers, the conformity of the goods sold to the established requirements systems, supplying them with a variety of safeguards to control the entire cycle of the consumer chain.

The successful development of the country's economic sphere and its integration with the global information space, calls for the activation of the development of modern telecommunications facilities and the expansion of the possibility of their use by all the participants of domestic and foreign trade turnover [18] in compliance with the basic legislative requirements for the digital economy.

References

1. ICT Market Report 2017/18: Preface and Table of Contents. https://www.eito.com (2018). Accessed 1 Jul 2018
2. Schlagwein, D., Conboy, K., Feller, J.: "Openness" with and without information technology: a framework and a brief history. J. Inf. Technol. **32**, 297–305 (2017)
3. Rubin, J., Samek, A., Sheremeta, R.M.: Loss aversion and the quantity-quality trade-off. Exp. Econ. 1–24 (2017)
4. The market for "Internet of things" and "Industrial Internet" in Russia and the World. http://www.directinfo.net (2018). Accessed 1 Jul 2018
5. Eversheim, W.: Innovation Management for Technical Products, p. 7. Springer-Verlag, Berlin, Heidelberg (2009)
6. Goi, C.-L.: The impact of technological innovation on building a sustainable city. Int. J. Qual. Innov. 3–6 (2017)
7. Yuce, B., Schaumont, P., Witteman, M.: Fault attacks on secure embedded software: threats, design, and evaluation. J. Hardw. Syst. Secur. 1–20 (2018) (Springer International Publishing AG, part of Springer Nature)
8. Yu, C.S., Tao, Y.H.: Understanding business-level innovation technology adoption. Technovation **29**(2), 92–109 (2009)
9. Internet Market Research In Russia. http://www.akit.ru (2018). Accessed 1 Jul 2018
10. Horch, A., Wohlfrom, A., Weisbecker, A.: An e-shop analysis with a focus on product data extraction. In: International Conference on Electronic Commerce and Web Technologies EC-Web 2016: E-Commerce and Web Technologies, pp. 61–72 (2017)

11. Chuprova, A.Y.: Criminally-legal mechanisms of regulation of relations in the sphere of electronic commerce, p. 14. D.Phil. thesis (2015)
12. Yokoo, M., Ito, T., Zhang, M., Lee, J., Matsuo T.: Electronic Commerce. Theory and Practice, p. 61. Springer-Verlag, Berlin, Heidelberg (2008)
13. Lopes, A., Valentim, N., Moraes, B., Zilse, R., Conte, T.: Applying user-centered techniques to analyze and design a mobile application. J. Softw. Eng. Res. Dev (2018)
14. Sklyarov, S.V., Evdokimov, K.N.: Modern approaches to the concept, structure and nature of computer crime in the Russian Federation. Criminol. J. Baikal Nat. Univ. Econ. Law **10**(2), 322–330 (2016)
15. Micklitz, H.-W., Pałka, P.: The empire strikes back: digital control of unfair terms of online services. J. Consum. Policy **40**(3), 367–388 (2017)
16. Miedema, T.E.: Consumer protection in cyber space and the ethics of stewardship. J. Consum. Policy 1–21 (2017)
17. Zimmerman, A., Chaudhry, P.: The Economics of Counterfeit Trade. Governments, Consumers, Pirates and Intellectual Property Rights, pp. 137–152. Springer-Verlag, Berlin, Heidelberg (2012)
18. Weber, R.H., Weber, R.: Internet of Things, pp. 101–125. Springer-Verlag, Berlin, Heidelberg (2010)

Part III
Counteraction of Terrorism and Extremism Challenges in Big Data-driven World

Counteracting the Spread of Socially Dangerous Information on the Internet: A Comparative Legal Study

Elena Yemelyanova, Ekaterina Khozikova, Anatoly Kononov
and Alla Opaleva

Abstract The research is devoted to the analysis of the Russian and foreign practices of countering the dissemination of socially dangerous information by means of the Internet, as well as the forms of the limitation of fundamental rights and freedoms to protect the foundations of the constitutional order, morality, health, rights and legitimate interests of others, to ensure the country's defense and state security. The purpose of this chapter is to analyze the phenomenon of socially dangerous information transmitted with the help of the Internet; to identify the essential features of the proliferation of negative content; to study the legal approaches and the world experience of restricting access to the Internet resources; to consider domestic and foreign practice of detecting and suppressing the dissemination of socially dangerous information by means of the Internet. The necessity of a complex and interdisciplinary approach to understanding the phenomenon of negative content is substantiated. Such approach will allow, in accordance with the modern scientific methodology of cognition, to construct a model explaining the specifics, structure, and dynamics of the dissemination of socially dangerous information in the context of its information-procedural conditioning. The necessity of proportionality of legal measures used for counteracting the threat of socially dangerous information and its real danger is proved, considering that the Internet is an information space symbolizing freedom of action and freedom of self-expression that are inherent in democratic states. The system of monitoring, searching, tracking and analyzing links on the Internet should not conflict with democratic ideals, such as freedom of speech,

---

E. Yemelyanova · E. Khozikova (✉) · A. Kononov
Academy of Management of the Ministry of Internal Affairs of Russia,
Moscow 125993, Russia
e-mail: khozikova@mail.ru

E. Yemelyanova
e-mail: eev-rusinovo@yandex.ru

A. Kononov
e-mail: amvd@mail.ru

A. Opaleva
Academy of the Prosecutor General's Office of Russia, Moscow 117638, Russia
e-mail: agp@agprf.org

© Springer Nature Switzerland AG 2019
A. G. Kravets (ed.), *Big Data-driven World: Legislation Issues and Control Technologies*, Studies in Systems, Decision and Control 181,
https://doi.org/10.1007/978-3-030-01358-5_13

open communication, and personal privacy. The implementation of any measures to improve the legal regulation of the counteraction to the dissemination of socially dangerous information by means of the Internet should be aimed at harmonization and unification of international norms in this field. Measures are proposed to improve the legal regulation of countering the dissemination of socially dangerous information transmitted through the Internet, as well as to harmonize international norms in this area.

Keywords Counteraction · Dissemination of socially dangerous information
The internet · Negative content
Ensuring rights and freedoms of a person and a citizen

1 Background

At present, the spread of socially dangerous information in the information and telecommunication "Interne" network (hereinafter - the Internet) has acquired a global character. The ubiquitous spread of the Internet leads to the commission of crimes both in the network itself, acting as the scene of the crime, and through the misuse of information, the dissemination of socially dangerous information or the so-called "negative content" [1].

At present, there is no universally recognized concept of socially dangerous information. Often this term is disclosed as an enumeration of various types of negative content.

Thus, V. N. Lopatin identifies harmful information as the information that provokes social, racial, national or religious hatred and animosity; calls for war; propaganda of hatred, enmity and superiority; distribution of pornography; infringement on honor, good name and business reputation of people; advertising (unfair, unreliable, unethical, obviously false, hidden); information that has a destructive effect on the psyche of people [2].

American sociologists J. Bryant and S. Thompson believe that negative content includes media violence, sexually explicit media content, threatening media products, extreme news, advertising cigarettes, alcohol and certain foodstuff harmful to health [3].

Without detracting from the merits of the above definitions, we note that in this study, the terms "socially dangerous information" and "negative content" will be used as equivalent ones.

An analysis of the phenomenon of socially dangerous information in the global Internet information network makes it possible to identify the characteristic subspecies of such information:

– propaganda of violence, hatred, and enmity based on race, national, religious or other socio-cultural affiliation;
– "sexually explicit content", including the spread of child pornography;

– propaganda of extremist and terrorist ideas, calls for the commission of terrorist acts, dissemination of information on the methods of committing terrorist acts and crimes of this category;
– the spread of national hatred and enmity (especially relevant in the VK ("V Kontakte") social network;
– propaganda of committing punishable crimes, demonstration of ways to commit them, and some other types of socially dangerous information.

The essential features of socially dangerous information in the Internet network include:

(1) accessibility, which is characterized by the fact that through the availability of the Internet, an unlimited number of people have access to it, especially young people and minors;
(2) the high social danger associated, as a rule, with the fact that the disseminated socially dangerous information inflicts damage or can cause damage to the interests of the person, society or state protected by law;
(3) negative impact on the personality of children and juveniles, who are special objects of legal protection;
(4) unlawfulness, connected both with the misuse of information networks and with the dissemination of socially dangerous information with their help;
(5) high latency [4];
(6) the difficulty or impossibility of identifying the perpetrators [5];
(7) distribution or positioning of marginal behavior among minors.

Considering these characteristics, it becomes obvious that counteracting the spread of socially dangerous information through the Internet is possible only by the joint efforts of the entire world community.

2 Analysis of the Current Legislation in the Field of Combating the Dissemination of Socially Dangerous Information

Nowadays, the protection of information and the restriction of the distribution of negative content is under the close attention of the entire world community. This is due both, to the development of international human rights law and the strengthening of the protection of individual categories of personal rights and freedoms in the national legislation of many states of the world.

Further, we'll consider in more detail some aspects of international legal regulation of the counteraction to the dissemination of socially dangerous information by means of the Internet.

The growing problem of spreading socially dangerous information through the Internet has led to the fact, that a symposium was held in Salzburg on topical problems of human rights and the media in September 1968. The result of the discussion of many topical issues was the adoption of the Declaration on the Media and Human Rights on January 23, 1970. The Part B of the Declaration provides for the prevention of dissemination of slanderous statements.

With the development of information and telecommunication technologies, the need to ensure the security of information is becoming increasingly important. Accordingly, the world community reacts to new types of crimes by adopting more advanced conventions and international normative legal acts affecting the current problems of the modern world.

The Convention on Cybercrime in the field of computer information, fixing the concepts of "computer system", "computer data", "service provider", does not identify signs of socially dangerous information.

Recommendation No. R (2001) 8 of the Committee of Ministers of the Council of Europe "On the issues of self-regulation of virtual content (self-regulation and protection of users against illegal or harmful content of new information and communication services)" indicates the need to identify a set of signs of harmful content of new communication and information services, for example, signs of content that includes violence and pornography, as well as of content encouraging the use of tobacco, alcohol or gambling, and of that content which allows anonymous and without supervision contacts between minors and adults. In this case, content providers should be encouraged to apply these content attributes to allow users to recognize and filter harmful content regardless of its origin.

The normative legal regulation of media content containing a demonstration of violence, cruelty or having a pornographic character is reflected in the Recommendations "On the principles of distribution of video-grams containing violence, cruelty or pornography" issued by the Committee of Ministers of the Council of Europe on 22.04.1989, No. R (89) 7 and in the Recommendations "On the depiction of violence in electronic media" dated 30.10.1997 No. R (97) 19 etc.

Taking into account the articles 8 and 10 of the "Convention for the Protection of Human Rights and Fundamental Freedoms" [6], Recommendation No. R (89) 7 of the Committee of Ministers of the Council of Europe to Member States "On the Principles for the Distribution of Video-grams Containing Violence, Cruelty or Pornography" emphasizes the importance of strengthening activities related to the distribution of video-grams containing violence, cruelty or pornography, as well as those that encourage drug abuse.

In particular, considering the protection of minors, it is recommended that the Governments of the participating States:

- should take measures to implement the principles outlined in the Recommendations;
- should ensure that these principles are known to all the persons and bodies concerned;
- should conduct a periodic evaluation of the effectiveness of the application of these principles in their internal law enforcement systems.

The Declaration on freedom of information exchange on the Internet contains the following provisions:

- the right to freedom of expression and information;
- the right to freedom of expression and the free exchange of information on the Internet;
- the balance between freedom of expression and information and other legitimate rights and interests;
- self-regulation of virtual information;
- concern about attempts to restrict public access to the exchange of information with the help of the Internet for political reasons or other reasons that are contrary to democratic principles;
- confidence in the need to firmly establish that the existing control over the exchange of information through the Internet, regardless of boundaries, should be carried out on an exceptional basis;
- the need to remove barriers to individual access to the Internet and complement the measures already taken to establish common access;
- guarantees of users' rights access to a wide range of information from a variety of national and foreign sources;
- the need to limit the liability of information provider services when they act as a simple transmitter or when they voluntarily provide access to or receive information from a third party;
- legal aspects of information public services, especially electronic trade in the virtual market;
- freedom of information exchange through the Internet should not infringe human dignity, rights and fundamental freedoms of other people, especially minors;
- maintaining a balance between respecting the wishes of the Internet user not to disclose his identity and the need for law enforcement agencies to identify those responsible for committing crimes;
- cooperation with law enforcement bodies in cases of the appearance of illegal information dissemination by means of the Internet.

In case that those responsible for the content of the information participate in an unjustified depiction of violence that grossly offends human dignity, harms the physical, mental or moral development of individuals, especially young people, exert negative emotional influence on the society, the member-states should effectively apply appropriate civil, criminal or administrative sanctions against them. The member states should promote the principle of non-violence in programs, services, and products that reflect the cultural diversity and wealth of the countries of Europe. The promotion must be conducted within the framework of various national and European programs aimed to support the production and distribution of audiovisual works and in close cooperation with European bodies and professional parties concerned [7].

Types of negative content are reflected in international treaties in the field of combating crime and terrorism. These include:

- child pornography (Convention "On Crime in the Sphere of Computer Information" of 23.11.2001);

- racist and xenophobic material, a motivated threat of racism and xenophobia, a racist and xenophobic motivated insult; negation, extreme minimization, approval or justification of genocide or crimes against humanity (Additional Protocol to the Convention "On Cybercrime in Relation to the Criminalization of Offences Related to the Manifestation of Racism and Xenophobia Committed by means of Computer Systems" of 28.01.2003);
- materials that promote crimes related to the sale of children, exploitation of children, child pornography, sexual abuse of children (Additional Protocol to the Convention "On the Rights of the Child in Relation to the Sale of Children, Child Prostitution and Child Pornography", Council of Europe "Convention on the Protection of Children against Sexual Exploitation and Sexual Abuse");
- shocking photographs or images of terrorist acts, hate stimulating statements (Recommendation No. 1706 (2005) of the Parliamentary Assembly of the Council of Europe "On Mass Media and Terrorism", Declaration "On Freedom of Expression and Information in the Media in the Context of Combating Terrorism").

Emphasizing the concern of the international community, it should be noted that some states also conduct a balanced work to counter the spread of socially dangerous information on the Internet. The consolidation of types of socially dangerous information transmitted with the help of the Internet in the legislation of foreign countries has its own peculiarities. Being participants in the international treaties reviewed above, (the fact which facilitates the unification of their national legislation regarding the legal regulation of negative content), the states introduce a certain variety in the process of legal regulation of negative content under the pressure of national specifics.

Negative content in the legislation of most Western countries (the US, the member states of the European Union) is divided into two categories: illegal and harmful, which differ in their legal net modes.

Illegal content is understood as legally prohibited information, the production and (or) distribution of which is a crime or an administrative offense. This view is taken by the International Association of "hot" Internet lines (INHOPE). Although sometimes the content prosecuted under the civil law is also identified as an illegal one.

Harmful content is understood as other socially dangerous information which is not prohibited by criminal or administrative legislation, but for which certain distribution restrictions are established. In both cases, the harm criteria for information remain different in the territory of different countries - each state decides itself what to consider legitimate and illegal on the basis of various cultural, moral, religious and historical norms and constitutional values.

The 2010 OSCE special study provides comparative data on the regulation of the Internet content by 56-member states of this organization [8]. At the same time, as noted in the report, most of the legal norms criminalizing the content are applicable to any means of communication and are not used exclusively for the Internet.

More detailed information, which covers different countries of the world (not only Western European ones), is contained in the analytical report "Content Filtering in the

Internet Network. Analysis of World Practice" [9]. It names the following categories of content that are filtered:

- political content (websites of opposition and human rights movements, religious movements and sects, information about ethnic minorities);
- socially dangerous information (pornography, LGBT propaganda, incitement of national and religious hatred, calls for violence);
- content related to national security (sites of terrorist and extremist organizations, resources of military adversaries, resources with confidential data, etc.);
- sites and services that violate economic interests (file-sharing sites and torrent trackers);
- specialized Internet tools and social services (search services, social networks, online translators).

This list also comprises other types of information, for example, information connected with the infringement of copyright, information of limited access.

In the Russian legislation, there is still no single classification of socially dangerous information. It can be distinguished only analytically, by studying individual legislative acts relating to various branches of law: constitutional, information, administrative and criminal.

The Constitution of the Russian Federation, including the right to freedom of thought and speech, the right to information and the freedom of mass information, establishes a legal prohibition against propaganda or agitation that engenders social, racial, national or religious hatred and enmity, as well as propaganda of the social, racial, national, religious or linguistic superiority. In addition, the Constitution of the Russian Federation allows for other forms of restriction of these rights and freedoms by federal law with a view to protecting the foundations of the constitutional order, morality, health, rights and legitimate interests of other persons, ensuring the country's defense and state security, while observing the criterion of proportionality of such restriction. The sectoral legal regulation of negative content in the Russian Federation is carried out on the base of this constitutional law [10].

The federal legislation also contains the rules governing the dissemination of information by means of the Internet. Thus, Federal Law No. 67-FL of 12.06.2002, prohibits calling for the commission of acts that fall under the definition of extremist activity, or otherwise inducing such acts, as well as calling for justifying extremism while conducting the pre-election campaign.

The fundamental source of the information law of Russia - Federal Law No. 149-FL of July 27, 2006, stipulates a general legal prohibition of the dissemination of information aimed at promoting war, incitement to national, racial or religious hatred and enmity, as well as other information the dissemination of which is prosecuted by criminal or administrative law.

The general logic of this prohibition implies that the information legislation provides prohibitions (restrictions) on the dissemination of information, the illegality of which is stipulated by criminal or administrative law (similar to the category of illegal content in foreign countries).

In addition, the Russian information legislation considers other types of negative information that are not provided for by the Criminal Code of the Russian Federation and the Code of Administrative Offences. It establishes the distribution restrictions in respect to these types of negative information (similar to the category of harmful content in Western states). For example, according to the Art. 15.1 of the Federal Law of July 27, 2006, No. 149-FL, the Russian information legislation prohibits not only materials with pornographic images of minors and/or information concerning the involvement of minors as performers for participation in pornographic entertainment events which is prosecuted by the criminal law (Art. 242.1 and 242.2 of the Criminal Code), but also the information concerning: a) ways and methods of development, manufacture and use of narcotic drugs, psychotropic substances and their precursors, means and places of cultivation of narcotic plants; b) methods of committing suicide, and also calls for committing suicide.

At the same time, the spread of the so-called "death groups", among which one of the most widespread and most dangerous groups is the "Blue Whale" group, has led to the fact that two criminal cases were initiated in the Chelyabinsk region in 2017. One of them relates to the accusation of committing crimes under three articles of the Criminal Code of the Russian Federation: Art. 110 in the old version ("Bringing to suicide"), Part 3 of the Art. 110.1 ("Inclination to committing suicide or facilitating the commission of suicide"), Part 2 of the Art. 110.2 ("Organization of activities aimed at encouraging the commission of suicide using the Internet"). The victims in this criminal case are two people—a 14-year-old girl from the Leninsky district of Chelyabinsk and an 18-year-old resident of the Amur Region [11].

Federal Law No. 436-FL of December 29, 2010, also implements the approach of self-identifying types of information harmful to children. The analyzed Law establishes two groups of information that are harmful to the health and (or) development of children: (1) prohibited for distribution among children, and (2) limited for distribution among children of certain age categories. These two types of information differ in terms of their degree of public danger and the legal modes of their dissemination.

These principles are implemented when conducting public events, performances, and the films by specifying age restrictions: +0, +6, +12, +16, +18.

3   Conclusions and Results

On the basis of the analysis, we can conclude that in the Russian Federation, as in most countries of the world, socially dangerous information which is disseminated using the Internet is divided into two types: forbidden and restrictedly prohibited. In general, this reflects the global trends implemented in the international conventions and treaties.

At the same time, the protection of an individual, especially minors, involves strengthening measures to control the spread of socially dangerous information by means of the Internet, which, in fact, is legally regulated in the modern world.

So, at the present time, we need to develop a complex, interdisciplinary approach to consider the phenomenon of negative content, which, in accordance with the modern scientific methodology of cognition, will allow to build a model explaining the specifics, structure and dynamics of the dissemination of socially dangerous information in the context of its information-procedural conditioning.

The necessity of proportionality of legal measures used for counteracting the threat of socially dangerous information and its real danger is proved, considering that the Internet is an information space symbolizing freedom of action and freedom of self-expression that are inherent in democratic states. The system of monitoring, searching, tracking and analyzing links on the Internet should not conflict with the democratic ideals, such as freedom of speech and open communication, the inviolability of the private sphere of life. The implementation of any measures to improve the legal regulation of the counteraction to the dissemination of socially dangerous information by means of the Internet should be aimed at harmonization and unification of international norms in this field.

References

1. Smirnov, A.A.: Negative content: identification problems in the context of legal regulation. Information Law **2**, 18–25 (2015)
2. Lopatin, V.N.: Information Security of Russia. Thesis for Doctor of Law. St. Petersburg. p. 433 (2000)
3. Bryant, J., Thompson, S.: Fundamentals of Media Exposure. McGraw-Hill Companies, (2002)
4. Perepelitsyn, V.A., Kravets, A.G.: The social networks' nodes grouping algorithm for the analysis of implicit communities. IISA 2016—7th International Conference on Information, Intelligence, Systems and Applications, art. no. 7785432, (2016)
5. Quyên, L.X., Kravets, A.G.: Development of a protocol to ensure the safety of user data in social networks, based on the Backes method. Commun. Comput. Inf. Sci. **466** CCIS, 393–399. (2014)
6. Convention "On the Protection of Human Rights and Fundamental Freedoms" concluded in Rome on 04 Nov 1950
7. Hasebrink, U., Livingstone, S., Haddon, L., Lafsson, K.: Comparing Children's On-Line Opportunities and Risks Across Europe: Cross-National Comparisons for EU Kids OnLine. LSE. London, EU Kids Online (2009)
8. Freedom of Expression on the Internet: Report of the OSCE representative D. Mijatovic. OSCE, Vienna (2011)
9. Content Filtering in the Internet Network. Analysis of World Practice": report of the Fund of the Development of Civil Society. (2013)
10. Constitution of the Russian Federation adopted by popular vote on 12 Dec 1993 (considering the amendments introduced by the laws of the Russian Federation on amendments to the constitution of the Russian Federation No. 6-FCL of 30 Dec 2008, No. 7-FCL of 30 Dec 2008, dated 05 Feb 2014, No. 2-FCL, of 21 July 2014, No. 11-FCL. RF Code, 2014 Aug 4, No. 31. Article 4398)
11. The official website of the Internet newspaper "Znak": https://www.znak.com/2017-09-19/e ksklyuzivnye_podrobnosti_ugolovnyh_del_o_kuratorah_suicidalnyh_grupp_v_chelyabinske (2017). Accessed 16 Oct 2017

Mechanisms of Countering the Dissemination of Extremist Materials on the Internet

Yury Latov, Leonid Grishchenko, Vladimir Gaponenko
and Fyodor Vasiliev

Abstract The security of society in connection with the increasing global informatization is changing in different directions and ambiguously: increasing in some aspects, it is significantly reduced in others. In particular, since the 1990s the active use of the Internet information network by terrorist and extremist organizations represents a real threat for many countries of the world, including the Russian Federation. The main methods of using the capabilities of information networks in extremist and terrorist activities are the promotion of this activity, the recruitment of new participants, and the communication of participants in this activity. The actualization of the problems of countering the use of information networks in extremist and terrorist activities led to the adoption in 2013–2014 a number of federal laws that significantly improve state control over the use of the Internet and promptly suppress its use for extremist and terrorist activities. The main practical problems are lack of special information and analytical support and a shortage of police officers with special knowledge and skills in information networks. First of all, it is necessary to implement automated systems for searching and monitoring sites on the Internet, the development of which is actively engaged in Roskomnadzor.

Keywords Cyber extremism · The internet · Crime counteraction
Legal regulation · Police activity in the Russian federation

Y. Latov · L. Grishchenko · V. Gaponenko (✉) · F. Vasiliev
Academy of Management of the Ministry of Internal Affairs of Russia,
Moscow 125993, Russia
e-mail: profgaponenko@gmail.com

Y. Latov
e-mail: latov@mail.ru

Y. Latov
Institute of Sociology FNISTS RAS, Moscow 125993, Russia

A. G. Kravets (ed.), *Big Data-driven World: Legislation Issues and Control
Technologies*, Studies in Systems, Decision and Control 181,
https://doi.org/10.1007/978-3-030-01358-5_14

145

1 Introduction

At least since the 1990s the enormous advantages of modern information and communication technologies and, above all, the global Internet, have been actively used in their own interests by groups and individuals who are engaged in extremist and terrorist activities [1]. Thanks to the Internet extremists and terrorists have received the widest opportunities to promote their ideology, influence the public consciousness effectively, attract supporters, provide reliable communications finance their activities, and obtain the necessary information for it. On the opinion of many experts (including law enforcement officials), the main propaganda of extremism and terrorism ideas is currently spread on the Internet, which poses a serious threat to the national security of many countries, including Russia [2–4]. At the same time, the fight against such phenomena is significantly hampered by the cross-border nature of the network and the peculiarities of network technologies. The number of subjects of terrorist and extremist activities using the Internet, cyber-terrorists, and cyber-extremists is constantly growing. According to the estimates of the Russian National Anti-Terrorist Committee (NAC), as early as in 2012, over 7000 extremist and terrorist websites functioned on the Internet, of which about 500 were in Russian. Current estimates are even higher. At the same time, a significant part of Russian-language sites related to cyber-terrorism and cyber-extremism is physically outside the Russian jurisdiction. The ideology of radical Islamism is most actively propagated in the Russian segment (as well as in other segments) of the network.

This phenomenon was called "cyber-jihad" (or "Internet Jihad"). A strong impetus to its development was given by the terrorist "Islamic state" [5, 6], whose leaders have repeatedly threatened the opening of military operations on the territory of Russia. Other extremist movements—nationalistic, neo-fascist, separatist, religious (the ideology of totalitarian sects) are also represented on the Internet. A significant part of such propaganda is connected with the centers of smoldering ethnic conflicts (first of all, in the republics of the North Caucasus).

The mechanism for using information networks by terrorist and extremist organizations represents a real threat to the Russian Federation. This threat began to be realized in the 2000s, and in the 2010s targeted measures were taken for its reduction and neutralization, that allowed to some extent to "curb" cyber extremism and cyber-terrorism.

The purpose of this analytical review is to characterize the main forms and methods of using information networks in extremist and terrorist activities, the review of cyber counter-extremism and cyber counter-terrorism legislative measures adopted in the Russian Federation in recent years, as well as recommendations for improving the effectiveness of counteraction cyber extremism and cyber-terrorism by Russian state bodies.

2 Characteristics of the Attractiveness of Information Networks for Cyber Extremists

The information network is traditionally defined in informatics as a system of communication between computers and other computing equipment intended for processing, storing and data transferring. In the Russian legislation the concept of "information and telecommunications network" is widely used, which is virtually equivalent to the concept of an information network operating on the physical basis of a communication network. According to Article 2 of the Federal Law dated June 27, 2006 "On information, information technologies and information protection" No. 149-FZ, an information, and telecommunication network is a technological system designed to transmit information on the communication lines, access to which is done using computer facilities. It appears that the concepts of "information network" and "information telecommunication network" can be used as equivalent.

The main trend of the recent decades is the integration of various previously operated separately networks on the basis of a single global information and telecommunication network Internet. According to various sociological surveys an active Internet audience in modern Russia (users who access the network at least once per day) exceeds 50% of the total population of the country, at least once a month about 2/3 of the population (i.e. practically all Russians, except children and the elderly). It is obvious that the Internet now is the most mass information network and for this reason is a convenient environment for various forms of criminal activity, including extremist and terrorist. Consider the features of the Internet, which determine its active use in extremist and terrorist activities.

2.1 Easiness in Receiving and Low Cost of Network Access

Currently, there are many options for connecting to the Internet, and the cost of access is constantly decreasing with the development of physical infrastructure (telecommunication networks). In most settlements of the Russian Federation, there are providers of network access services offering a permanent (wired) high-speed broadband connection. Tariffs of providers for residential premises of individuals are constantly decreasing and with an average payment level of about 500 rubles per month can reach the minimum values of 100–200 rubles for the monthly service.

In addition, most of the settlements, as well as a significant part of the territories outside the settlements, are currently in the coverage area of cellular networks providing wireless Internet access services, including high-speed broadband. The services of cellular operators are more expensive and are usually provided with limited amounts of information transmitted or access speed, however, in most cases, there are connection options with a cost of 300–400 rubles per month or even cheaper. In addition to cellular networks, a number of operators of wireless access networks, built on other technologies (Wi-Fi, etc.) operate in a number of cities.

Finally, in settlements, especially large ones, there are numerous options for free access to the Internet through numerous Wi-Fi access points installed in public places, in food enterprises, transport facilities (including the metro), etc. Free access to the network is often provided in educational organizations and libraries.

Thus, now in Russia, almost anyone can easily get inexpensive or even completely free access to Internet resources.

2.2 The Potentially Unlimited Audience in the Country and the World for the Dissemination of Information

The capacity of network resources to influence public consciousness is comparable to that of traditional media, or, according to a number of studies, it is already exceeding it. This influence is especially strong on the younger generation of citizens, whose representatives prefer to receive information not from newspapers, radio and television programs, but from network sources. The sources of information for them are not only electronic media but also social networks and blogs.

By carrying out information activities in the network, extremist and terrorist organizations and their accomplices get a virtually unlimited audience. This is due to the fact that they have the opportunity to reach the audience and traditional media, which are increasingly using the Internet as a source for their messages. Thus, the use of Internet resources in principle gives an additional opportunity to influence an audience which for some reason does not have access to the network.

2.3 Insufficient State Control

Due to the peculiarities of the Internet network, its cross-border and international character, most states have limited opportunities to regulate and control the use of network resources. The greatest number of levers of influence on network resources is concentrated in the hands of the creators of the Internet network—American government bodies and non-profit organizations.

Most protocols and Internet technologies are developed in the USA. The Corporation for the Management of Domain Names and IP-addresses (ICANN) was established with the participation of the US government and is located in the United States. Most of the root DNS servers containing information about top-level domains and support the operation of these domains are also located in the US or Western Europe. Accordingly, information about users of the listed systems is at the disposal of American business companies. Obtaining information from them for Russian law enforcement agencies, unfortunately, is a difficult task, most often impossible.

The political contradictions between Russia and Western countries, which intensified in the 2010s, introduce additional difficulties since extremists often impersonate persecuted dissidents and receive support from the West. A striking example is a site of the Chechen separatists Kavkaz-Center, previously successfully working on European hosting platforms, and now safely moved to San Francisco.

2.4 High Anonymity, Security, and Secrecy of Communication and Information Exchange on the Internet

There is a significant number of different technologies and technical solutions that allow anonymity of the use of Internet resources, as well as protect information from being intercepted over the network. Most of these technical solutions are either completely free of charge or very inexpensive and are available to users who do not have any special knowledge of computer technology.

To this end, various special technologies and systems are used, many of which are currently available. First of all, it is a VPN (Virtual Private Network), a group of technologies that allow a secure network connection (logical network) over the Internet. Usually, VPN is provided as a paid service by numerous official and unofficial suppliers, but the most technically advanced criminal groups create their own "private" VPN servers.

Another widely used method of hiding real IP addresses is the use of so-called anonymous networks working on the Internet and specifically designed to ensure the anonymity of connections. In such networks, traffic encryption and the distributed structure of network nodes are used. The most popular anonymous network is TOR, and there are several dozens of them all.

2.5 High Dynamic Network Resources

The procedures for creating a website, blog, individual account or group on a social network, registering an electronic mailbox or other means of electronic communication on the Internet are extremely simplified, requiring minimal skills and little time, effort and expense. Financial investments are necessary (and not always) only when creating a separate site—to pay for domain name registration and hosting service, but the cost of such services, as a rule, is relatively low. This makes it easy to create resources and use any type of network communication, and it's also easy to replace them if necessary.

As noted, for example, in the report of Gabriel Weimann, "terrorism in the Internet is a very dynamic phenomenon: websites appear suddenly, they often change the format, and then as quickly disappear—or, in many cases, create the appearance of disappearance, changing their address, but keeping the content" [7]. Such dynamism of network technologies allows participants of extremist and criminal activity to create information resources and use network communication facilities faster than law enforcement bodies manage to adequately react to it.

2.6   Using Multimedia Capabilities

The information on the Internet is distributed in the form of texts, graphic images, audio and video records, etc., which is actively used in the interests of extremist and terrorist activities, primarily in order to promote their activities and influence the mass audience.

Multimedia, audiovisual materials exert a much stronger influence, especially on young people, who are characterized by so-called "mosaic thinking"—the perception of information through short, bright and simple images. The way of providing information in the format of short videos is especially popular now and this format is actively used by representatives of extremist and terrorist groups around the world.

3   The Main Directions for Using the Capabilities of Information Networks in Extremist Activities

The use of the Internet opens new vast opportunities for extremist and terrorist groups:

- to carry out propaganda of their activities, affecting a significant number of recipients of information;
- to find and recruit new participants;
- receive training materials, as well as detailed intelligence information on practically any object, organization or person;
- finally, commit acts of "cyber terrorism" and other crimes.

Let us consider in more details the main forms and methods of using the capabilities of information networks in extremist activities.

3.1   Propaganda of Extremist and Terrorist Activities

Propaganda is the most common form of using the capabilities of information networks in terrorist and extremist activities. The main sources of propaganda by extremist and terrorist organizations are both the own resources of extremist and terrorist organizations, as well as any other network resources—official Internet media, websites of individuals and organizations, forums of various orientations, blogs, social networks, video hosting, etc., on which it is possible to post materials or comment on previously placed materials.

Open propaganda usually consists of the direct glorification and approval of illegal and violent acts committed by extremists and terrorists, accompanied by calls to commit such acts, and also to join extremist and terrorist groups. Propaganda can also be indirect, the goal of which is to maintain certain attitudes among the audience.

In their own way, outstanding achievements in this regard are demonstrated by representatives of the "Islamic state" who are characterized not only by their phenomenal cruelty but also by the active propaganda of their ideas in networks. Propaganda is structured in such a way that appeals in support of the actions of this group are even made by the sentenced to execution victims just before the killings. This is carefully recorded on video recordings which are then distributed as widely as possible.

Propaganda of extremist and terrorist activities on the Internet depending on the form and direction can be qualified as crimes under Art. 205.2 of the Criminal Code of the Russian Federation "Public appeals for the implementation of terrorist activities or the public justification of terrorism", Art. 280 of the Criminal Code of the Russian Federation "Public calls for the implementation of extremist activities", Art. 280.1 of the Criminal Code of the Russian Federation "Public calls for the implementation of actions aimed at violating the territorial integrity of the Russian Federation", Art. 282 of the Criminal Code of the Russian Federation "Raising hatred or enmity as well as the humiliation of human dignity".

The very first verdict for cyber-extremism was made in Russia more than 10 years ago and is connected with propaganda. This is the verdict made in 2006 for distributing information of an extremist nature on the Internet: a student of the Faculty of Philology and Journalism, gr. K. posted a text on his website containing propaganda of the activities of the Kemerovo national Bolsheviks as well as calls for a violent change of the political system in the Russian Federation [8].

A more recent and well-known example of the criminal prosecution for the propaganda of terrorism on the Internet is the initiation and investigation of a criminal case against the leader of the Ukrainian radical group Right Sector D. A. Yarosh. On March 3, 2014, the Criminal Code of the Russian Federation initiated a criminal case against him under Part 2 of Art. 205.2 and Part 2 of Art. 280 of the Criminal Code of the Russian Federation on the fact of placing in February 2014 on the page of the "Right Sector" in the social network "In Contact" an appeal addressed to the leader of Chechen terrorists Doku Umarov to commit terrorist acts on the territory of Russia. As a result, in this case, D. A. Yarosh was declared internationally wanted by Interpol although for political reasons this criminal prosecution was not developed.

3.2 Recruiting New Member of Extremist and Terrorist Activities

The attraction of new participants in extremist and terrorist activities can be identified as a separate, especially dangerous form of using information networks by extremists and terrorists. A large audience of the Internet provides groups and their supporters with a large pool of potential recruits.

Recruitment can be carried out in the form of direct appeals to participate in extremist or terrorist activities placed on the own resources of the relevant group- ings or on any other resources where their open propaganda is conducted. A more complicated form of recruitment is a hidden search of persons among the visitors of resources or participants of the discussions who are of particular interest to extrem- ists and terrorists as potential followers, accomplices or performers. Recruiters can use any network technology; navigate through social networks and forums in search of the most receptive audience, especially the youth. Individual processing of found candidates can be conducted already through closed communication channels.

The recruitment of new members of extremist and terrorist activities on the Internet can be qualified as crimes under Art. 205.1 of the Criminal Code of the Russian Federation "Assistance to terrorist activities" ("Declination, recruitment or other involvement of a person in committing at least one of the crimes provided for in Articles 205, 206, 208, 211, 277, 278, 279 and 360 of this Code") or Part 1.1 of Art. 282.1 of the Criminal Code of the Russian Federation "Organization of an extremist community" ("Declination, recruitment or other involvement of a person in the activities of an extremist community").

An example of a criminal case on the fact of recruitment is a case initiated by the FSB for the Astrakhan region in September 2014 under art. 205.1 in respect of the local resident. The figurant recruited new members to "The Mujahedeen Shama" grouping by posting a video on the YouTube service with calls for support for jihad. The fact that the video was posted was identified by the police. At the request of the prosecutor's office, the video was found extremist, removed from network resources, and the person was put on the wanted list.

3.3 Information Support for Extremist and Terrorist Activities

The global Internet is a universal source of information. In extremist and terrorist activities information received from the network is used mainly in two directions—for training participants in criminal activities and for collecting information about objects of criminal influence. Counteraction to the dissemination of extremist materials is connected with the first direction of cyber extremism.

A significant number of sites on the Internet contain materials about the creation of various weapons, explosives, hazardous chemicals, detailed instructions for com- mitting various crimes and countering law enforcement. Classical examples are the books "The Guide to the Terrorist" and "Anarchist Cookbook" devoted to a detailed description of the process of creating various types of explosive devices. A volu- minous guide entitled "The Jihad Encyclopedia", prepared by Al-Qaeda, contains detailed instructions on how to create an underground organization and conduct attacks.

A separate type of teaching materials widely used at the present time is videos showing the processes of preparation for crimes and their commission. Numerous teaching materials also give detailed instructions for organizing secure communications, encrypting messages using anonymization methods. The use of interactive and multimedia technologies helps to create a sense of community between people living in different regions and of different origins and facilitates the creation of networks for the exchange of educational and tactical materials.

Such materials are used by extremist and terrorist groups to train and improve the skills of their participants, and also fall into the hands of people with unstable minds committing single crimes. For example, in Finland in 2002, a 19-year-old chemist Petri Gerdt studied and discussed methods of manufacturing explosives for a while on the web forum of chemists "Kotikemia" (Home Chemistry). Based on the information received during the discussion he produced an improvised explosive device and on October 11, 2002, detonated it in a shopping center in Vantaa, killing seven people, including Gerdt himself. The forum where he received the information was closed by the sponsoring company after the incident as it turned out that it contained a large amount of information about the manufacture of explosives. However, a copy intended only for reading was immediately posted on the network [9].

Training with the use of network technologies and materials from information networks for the purpose of carrying out terrorist activities can be qualified under art. 205.3 of the RF Criminal Code "Training for the purpose of carrying out terrorist activities". According to this norm a person's training is considered as such if knowingly conducted for the purpose of carrying out terrorist activities or committing one of the crimes provided for in articles 205.1, 206, 208, 211, 277, 278, 279 and 360 of the RF Criminal Code, including practical skills in the course of physical and psychological training when studying the methods of committing these crimes, the rules of handling weapons, explosive devices, explosive, poisonous, as well as other substances and subjects representing danger to others.

If training for the purpose of carrying out terrorist activities is not conducted independently but by a certain person, a "mentor", art. 205.1 of the RF Criminal Code "Assistance to terrorist activities" which establishes punishment, in particular, for the preparation of a person for the purpose of committing at least one of the crimes provided for in Art. 205, 206, 208, 211, 277, 278, 279 and 360 of the RF Criminal Code, can be applied to his actions.

Training or preparation for the purpose of carrying out extremist activities as separate crimes is not provided in the RF Criminal Code; if, however, such training may be one of the signs of organizing a criminal community qualified under art. 282.1 of the RF Criminal Code "Organization of an extremist community" (in particular, "the creation of an extremist community, that is, an organized group of individuals for the preparation or commission of extremist crimes").

4 Formation of the Russian Legislation to Counteract the Use of Information Networks in Extremist and Terrorist Activities

As already indicated, the cooperation of the Russian law enforcement agencies with foreign states and companies in combating extremist and terrorist activities in information networks continues to be problematic.

One of the main reasons for this (apart from the political contradictions between Russia and the West) is the lack of unified legal approaches to the criminalization of acts. It should be noted that in the criminal legislation of most states there are no concepts of extremism and extremist activity, and only a few specific acts are provided, such as the incitement of national, racial and religious hatred; insults, threats and violence on the grounds of belonging to any groups; the spread of Nazi and racist ideology and literature; the use of appropriate symbols, etc. Such concepts as "discrimination", "xenophobia", "anti-Semitism", "Islam phobia", "crimes committed out of hatred" (hated crime) are also used. Struggle against such crimes is given increased attention in all developed countries as there is a clear-cut upward dynamics of their number everywhere.

However, up to the present moment, there is no comprehensive international treaty on combating terrorism and extremist crimes that would apply to the widest possible list of their manifestations. Since 2000, the UN member states have been negotiating to conclude a comprehensive convention on the suppression of terrorism which will ultimately include the definition of terrorism. Of course, there is no universal convention specifically aimed at preventing and suppressing the use of information networks by terrorists and extremists, especially the Internet.

The only legally binding multilateral document on combating criminal activities carried out using the Internet is currently the Council of Europe's Convention on Computer Crime, adopted in 2001 (the so-called Budapest Convention). This Convention contains a classification of computer crimes and details the mechanisms for the interaction of law enforcement agencies of individual states in situations where the offender and victim are in different countries. In 2002, in addition to the Convention, Protocol No. 1 was adopted, adding to the list of crimes the dissemination of information of a racist and other nature inciting to violence, hatred or discrimination of an individual or a group of persons based on race, nationality, religion or ethnicity. The Convention also specifies for all operators common rules for storing information about customers in case such information is required in the investigation of crimes. According to one of the provisions of the Convention, "a Party may, without the consent of the other Party, obtain on its territory through the computer system an access to the data stored on the territory of the other Party or obtain it if this Party has the legal and voluntary consent of a person who has the legal authority to disclose this data to this Party through such a computer system". Because of this provision, which allows foreign special services directly request information from operators of other states, the Russian Federation does not plan to join the convention. Thus, Russia's interaction with non-CIS countries is currently complicated on the whole not only by the insufficiency of the existing treaty base but also by the foreign policy problems that have arisen recently. Thus, Russia's interaction with non-CIS countries is currently complicated on the whole not only by the inadequacy of the existing treaty base but also by the foreign policy problems that have arisen recently.

Unable to actively develop international cooperation in countering cyber extremism, the Russian state has taken independent actions in recent years aimed at countering extremist and terrorist activities using information networks. The most important legislative acts were adopted in 2013–2014.

The intensification of counteraction cyber extremism is connected, first of all, with Federal Law No. 398-FZ of December 28, 2013 "On Amendments to the Federal Law" "On information, information technologies, and Information protection", adopted five years ago (the so-called "law of Lugovoi"). According to it, Article 15.3 "The procedure for restricting access to the information disseminated in contravention of the law." was added to the Federal Law "On information, information technologies, and information protection". This provision allows the RF General Prosecutor and his deputies, in case of detection in the information and telecommunication networks, including the Internet, of the above mentioned information to send to the Federal Service for supervision in the sphere of communication, information technology and mass communications of Russia (Roskomnadzor) requirements for taking measures to limit access to information resources.

Roskomnadzor takes the following measures:

1. Directs on the "information system of interaction" (the site of the system—http://vigruzki.rkn.gov.ru) to telecommunication operators providing Internet access services on the territory of the Russian Federation; the requirement to take measures to restrict access to the information resource, including a website on the Internet, or to information posted on it. This requirement should contain the domain name of the site, the network address, the indexes of the pages of the site on the network, allowing to identify such information. Under the requirement the interaction occurs mainly in the automatic mode; communication operators block access to resources promptly. Thus, almost immediately after sending a requirement from the entire territory of the Russian Federation it becomes impossible to access the corresponding site or its separate page. To circumvent the restriction is possible only with the use of means of anonymization and hiding access (VPN, proxy, etc.) which is not used by most of the population.

2. Identifies a hosting provider or another person who ensures the placement of a resource in the information and telecommunications network; this resource serves the owner of the site on which the information is posted. A provider or another person is notified electronically in Russian and English about the violation of the procedure for the dissemination of information indicating the domain name and network address that allow to identify the site on the Internet and the requirement to take measures to delete information. In other words, the requirement t is not sent to the person who posted the information and not the site owner, but to an organization or a person providing the work (hosting) of the relevant site. Within one day from the receipt of the notification, the

hosting provider is obliged to inform the site owner and notify him of the need to promptly delete the information containing calls for riots, extremist activities, participation in mass (public) events conducted in violation of the public order. If the owner removes the information and reports this to Roskomnadzor access to the site is restored by transferring the appropriate instruction to the communication operators through the interaction system.

An important feature of this procedure is that the RF General Prosecutor and his deputies may demand the restriction of the access to the sites on the basis of receipt of notifications from federal state authorities, state authorities of the subjects of the Russian Federation, local governments, organizations or citizens. Thus, it is possible for counter-extremism units of the Ministry of Internal Affairs of Russia to send notifications to the General Prosecutor of the Russian Federation in the course of solving their operational and official tasks.

This order of access restriction is most effective with respect to the resources hosting of which is carried out in the territory of the Russian Federation, but the owners of foreign websites, especially large ones, in some cases will have to comply with the requirements of the Russian legislation in order not to lose the flow of visitors from Russia. The new order came into force on February 1, 2014, and by September 2014, according to the report of Roskomnadzor head A. A. Zharov, the access to more than 600 different sites on the Internet was blocked taking into account the new order. It's necessary to compare the official figure for 2017 when, according to the report of the National Anti-Terrorist Committee (NAC) head Alexander Bortnikov, the operation of over 2 thousand terrorist and extremist resources was already suppressed.

The next important link in the improvement of the legislation aimed at combating extremist and terrorist crimes committed with the use of information networks was the adoption of the so-called "law on bloggers". The Federal Law "On information, information technologies and information protection" and certain legislative acts of the Russian Federation on the regulation of information exchange with the use of information and telecommunications networks obligated the authors (owners) of Internet resources (sites, blogs, etc.) with the audience of over 3000 users per day to register in Roskomnadzor and imposed a number of restrictions on the content of these resources. Owners of popular resources, including sites, pages, accounts in social networks and blogs, were required to register them in Roskomnadzor and comply with Russian legislation regardless of resources profiles, placements, and domain zones.

The owners of websites and pages on the Internet began to apply all the rules established for the media. For violation of the established requirements, administrative liability is provided. This very law added Art. 13.31 "Failure to fulfill the duties of the organizer of the dissemination of information on the Internet" and Art. 19.7.10 "Failure to provide information or submission of knowingly unreliable information to the body exercising control and supervision functions in the field of communications, information technology, and mass communications" in the Code of Administrative Offenses of the Russian Federation. In case the organizer of the dissemination of information on the Internet does not fulfill the established obligations, restriction of access to the information resource of the organizer is applied.

From the point of view of combating the use of information networks in extremist and terrorist activities the "law on bloggers" provided a convenient mechanism for identifying the owners of popular sites and pages of any type and providing them with the necessary legal impact.

The Federal Law No. 178-FZ of June 28, 2014, "On amending certain legislative acts of the Russian Federation" was also important for countering cybercrime; the law not only made amendments but also refined the wording of articles 280 and 282 of the Criminal Code. The crime commission with the use of "information and telecommunication networks including the Internet" was established as a possible way of committing crimes. Thus, the qualification of extremist crimes most frequently committed with the use of the Internet was simplified. It was also possible to solve the problem of qualification of crimes under art. 282 of the Criminal Code of the Russian Federation since earlier in the cases of their commission using the resources of the network, only the sign of committing "publicly" or "using the mass media" was approached. The difficulty in practice was caused by the fact that not all actions performed on the network are of a public nature (for example, correspondence by e-mail) and only a few resources were officially classified as media.

Therefore, in 2013–2014 there was a kind of turning point in the fight against cyber extremism in Russia. It was connected with the adoption of a number of legislative acts directly or indirectly aimed at combating the use of information networks in extremist and terrorist activities. This policy of the state aimed at limiting the uncontrolled use of information networks was generally approved by the population of Russia: in particular, according to a sociological survey conducted by the Levada Center, at the end of 2014 about 54% of respondents believed that in the Internet "there are many dangerous sites and materials", so their censorship is necessary.

The aggregate application of new rules of law generally improved the state of legality in the Russian-language segment of the Internet and led to a reduction in the most dangerous manifestations of cyber-extremist and cyber-terrorist activities. The measures are taken partially compensate for shortcomings in the sphere of international cooperation which due to political circumstances will probably be observed for quite a long time.

5   Problems of Improving the Organization of Countering the Use of Information Networks in Extremist and Terrorist Activities

Fracture of 2013–2014 in the legal support of the fight against cybercrime found a positive response from the staff of the Russian law enforcement agencies.

Sociological surveys conducted in 2015 among the students of the Academy of Management of the Ministry of Internal Affairs of Russia representing the entire range of units of the Ministry of Internal Affairs of Russia showed that the Ministry of Internal Affairs employees fully understand the need to counter extremist and terrorist activities on the Internet.

The results achieved in this regard are estimated positively by approximately 65% of the respondents. At the same time, a comparatively high evaluation of the measures used is made: the compilation of "black" lists is considered an effective measure to some extent (the answers are "yes" and "sooner, yes") by 72% of respondents; the closure of extremist sites—77%, criminal prosecution—67%.

At the same time the police officers are generally satisfied with the measures applied, or at least they do not see an effective alternative for them. At the same time, the majority of respondents believed that only about 30–35% of crimes related to the distribution of extremist materials on the Internet can be identified and detected, thus marking their high latency.

One of the reasons for this is the lack of resources to counteract extremist activities on the Internet. To ensure that this counteraction was not potential, but really effective, three key resources are needed: normative powers, police qualifications and logistical support. The answers of the respondents showed that there is a deficit for each of these resources. In the area of security of authority (39% of respondents give a positive assessment, 34%—negative), the situation is better than in the provision of qualified employees (29% gave a positive assessment, 45%—a negative one).

An important role is played by the problem of interaction between units for countering extremism and the Federal Service for supervision of communications, information technology and mass communications of Russia (Roskomnadzor) and its territorial subdivisions. According to Article 15.3 of the Federal Law "On information, information technologies, and information protection" Roskomnadzor received broad opportunities to effectively influence resources on the Internet, and in case of refusal to comply with the requirements of the legislation by its owners to restrict access to resources from the territory of Russia. The problem is who and how will organize counter-extremist Internet monitoring.

For the independent implementation of supervisory powers, Roskomnadzor develops an automated system for searching for various information on the Internet, the distribution of which is prohibited or restricted by the legislation of the Russian Federation. The system will perform the automated intellectual search for keywords and phrases, and its results will be monitored by employees of a specially created large unit within Fedearal State Enterprise "GRCTS". The developed methodology of using the system presupposes its application among other tasks and for the identification of materials of an extremist and terrorist nature.

The main components of information and analytical work in the fight against the use of information networks by extremists and terrorists are the automation of monitoring of network resources, as well as the formation and analysis of specialized accounts (databases).

Automation of the Internet monitoring in order to identify the use of resources in the interests of extremist and terrorist activities is still an unsolved problem of the units to counter extremism of the Russian Ministry of Internal Affairs. At the moment, they do not have any automated systems for searching and monitoring sites on the Internet, and so this work is carried out personally by the employees using standard technical means by searching for information with public means and by visually inspecting the pages of the sites. This work is labor-intensive and requires a lot of time, so the task of developing and using special search and analytical systems is relevant and promising. The search and tracing of manifestations of extremist and terrorist activity by detectives of the units for counteracting extremism "manually" with the use of standard personal computers certainly have serious limitations on speed, coverage, and efficiency.

The automation of monitoring could solve the problem of:

- identification of previously unknown resources on which the placement of materials of an extremist and terrorist nature is conducted on a regular basis or communication of participants of criminal activity is conducted;
- search for extremist and terrorist materials, including texts, images, and videos;
- tracking the facts of placing previously known extremist and terrorist materials on new resources;
- tracking the activity of people on the network permanently placing materials of an extremist and terrorist nature identified by pseudonyms, e-mail addresses, other contact information and individual network characteristics;
- establishment of implicit links between persons placing materials of an extremist or terrorist nature or participating in any other way in the work of identified resources;
- preservation of the full copies of the pages where the materials are placed in the format necessary for the subsequent compilation of operational and the official documents, as well as for the purpose of saving in case the materials are deleted from the pages of the sites.

Currently, there are a number of hardware and software solutions designed for automatic counter-extremist monitoring developed both by foreign [10] and Russian specialists. The selection and optimization of any of the proposed systems for the needs of units to combat extremism of the Russian Ministry of Internal Affairs or the development of a new system is a complex of the scientific and practical task that requires preliminary working out and adoption of a number of managerial and personnel decisions.

Another urgent task of organizing effective information and analytical work of units to counter extremism is the creation of a database for storing and processing operational and reference information obtained both as a result of monitoring the resources of the Internet (in the future—automatically) and from other sources, including operational. It is advisable to store exhaustive information about identified individuals and organizations involved in extremist and terrorist activities, unidentified persons who are active on the Internet (identified and tracked by some signs) and in real life, about all extremist and terrorist materials. The database can be operated by the units for countering extremism independently or operate in the system of information records of operational intelligence units of the Ministry of Internal Affairs of Russia.

6 Conclusion

Analysis of the issues related to the development of cyber-extremism and counter-action to it allows us to formulate the following conclusions about the main trends in this area in Russia in the 2010s.

1. At present, the vast majority of adult citizens of Russia is active users of information networks and is potentially exposed to various forms of cyber-extremist activity related to the use of these networks. Such impact is facilitated by a number of objective characteristics of the development of information networks—easiness of access to networks, their wide audience, weak state control over networks, anonymity in the network and a wide range of information transfer opportunities over networks.
2. The main methods of using the capabilities of information networks in extremist activities are the promotion of this activity, the recruitment of new participants, the financial and information support of such activities, communication of participants in this activity. Most of these acts fall under the various articles of the Criminal Code of the Russian Federation; in the last decade, there are already many examples of initiating criminal cases against cyber extremists.
3. The actualization of the problems of countering the use of information networks in extremist and terrorist activities led to the adoption in 2013–2014. a number of long-awaited federal laws that allow to significantly increase state control over the use of the Internet and promptly suppress its use for extremist and terrorist activities. At the same time, at the international level, the normative legal support in this sphere remains weak and hampers international cooperation in this area.
4. "Internetization" of extremist and terrorist activities requires police officers to search for new organizational ways to improve the counteraction to extremism and terrorism. The main practical problems are the lack of special information and analytical support and the shortage of employees with special knowledge and skills in the information networks. First of all, it is necessary to implement automated systems for searching and monitoring sites on the Internet.

Summarizing, it should be emphasized that the emerging international situation, the internal political situation and the development trends of information and telecommunication technologies make it possible to predict further growth of the use of information networks in extremist and terrorist activities. In particular, there is an objective danger of the widescale use of the possibilities of modern information technologies in the preparation for mass riots and individual extremist crimes of an extremist and terrorist nature (such as "Internet Jihad"). Timely prevention and suppression of existing threats depend on the effectiveness, consistency, and coherence of activities conducted by law enforcement and other state bodies and aimed at countering the use of information networks in extremist and terrorist activities. At the same time, preventing the criminal misuse of information networks requires scientifically based measurement of benefits and costs so that combating crime does not create high barriers to the legitimate use of information networks.

References

1. Whine, M.: Cyberspace-a new medium for communication, command, and control by extremists. Stud. Confl. Terrorism **22**(3), 231–246 (1999)
2. Chen, H.: Exploring Extremism and Terrorism on the Web: The Dark Web Project Pacific-Asia Workshop on Intelligence and Security Informatics. Intelligence and Security Informatics, pp. 1–20 (2007)
3. Denning, D.: Terror's web: how the internet is transforming terrorism In: M. Yar, & Y. Jewekes (Eds.), Handbook of Internet Crime, Willan Publishers, pp. 194–212 (2010)
4. Awan, I.: Debating the meaning of cyber terrorism: issues and problems Internet journal of criminology, pp. 1–10 (2013)
5. Klausen, J.: Tweeting the Jihad: social media networks of western foreign fighters in Syria and Iraq. Stud. Confl. Terrorism **38**, 1–22 (2015)
6. Awan, I.: Cyber-extremism: Isis and the power of social media society, April, **54**(2), 138–149. https://link.springer.com/article/10.1007/s12115-017-0114-0 (2017). Accessed 1 Jul 2018
7. Weimann, G.: How Modern Terrorism Uses the Internet. Special Report 116. http://www.usip.org/sites/default/files/sr116.pdf (2004). Accessed 1 Jul 2018
8. Borisov, S.V., Vasnetsova, A.S., Zhafyarov, A.G.: On the issue of countering cyber terrorism and cyber extremism. Bulletin of the Academy of the General Prosecutor's Office of the Russian Federation. **1**, 52 https://elibrary.ru/download/elibrary_24902312_82681100.pdf (2015). Accessed 1 Jul 2018
9. Myyrmanni bombing. https://fi.wikipedia.org/wiki/Myyrmannin_räjähdys (2018). Accessed 1 Jul 2018
10. Scanlon, J.R., Gerber, M.S.: Automatic detection of cyber-recruitment by violent extremists. Security Informatics. https://link.springer.com/article/10.1186/s13388-014-0005-5 (2014). Accessed 1 Jul 2018

Analysis of High-Technology Mechanisms of Extremist and Terrorist Activities Financing

Boris Borin, Irina Mozhaeva, Valery Elinsky and Oleg Levchenko

Abstract Modern political and legal processes of globalization, expansion of network structures and organizations are intensifying the development of extremist and terrorist activities and the process of their multi-level financing. In this regard, it is relevant not only to study the problems of countering extremist and terrorist activities but also to analyze the dependence of extremist and terrorist activities on their level of funding. The authors examine the legal basis for combating the financing of extremism and terrorism; characterization of the ways of organizing financing of extremist and terrorist activities; types of sources and channels for the financing of these criminal activities. The identification of methods, as well as sources and channels for financing terrorism and extremism, will facilitate the development of an algorithm of law enforcement actions to identify and interdict them in order to bring the perpetrators to criminal liability. The findings revealed in the course of the study indicate that the most important direction in counteracting terrorist and extremist activities is to neutralize the ways of financing them and eliminate sources of criminal proceeds, as well as to optimize the forces and means to counter terrorist, extremist activities.

Keywords Law enforcement agencies · Extremist and terrorist crimes
Activities of terrorist and extremist organizations
Financing of extremist and terrorist activities
Sources of terrorism and extremism financing

B. Borin · I. Mozhaeva (✉) · V. Elinsky · O. Levchenko
Academy of Management of the Ministry of Internal Affairs of Russia,
Moscow 125993, Russia
e-mail: mirina-crim@yandex.ru

B. Borin
e-mail: bvb007@yandex.ru

V. Elinsky
e-mail: volakdm@va-mvd.ru

O. Levchenko
e-mail: katolean@gmail.com

© Springer Nature Switzerland AG 2019
A. G. Kravets (ed.), *Big Data-driven World: Legislation Issues and Control Technologies*, Studies in Systems, Decision and Control 181,
https://doi.org/10.1007/978-3-030-01358-5_15

1 Introduction

In modern conditions, the activities of terrorist and extremist organizations have assumed the transnational nature, and pose a real threat to the security of both individual states and the international community as a whole. One of the main sources of threats to the national security of the Russian Federation is the activities of terrorist and extremist organizations, as well as radical public associations, non-profit non-governmental organizations, financial and economic organizations and private individuals who provide assistance, including financing, in the conduct of terrorist activities [1].

The object of the study is social relations, formed in the process of countering extremist and terrorist activities. The subject of the study is the legal and organizational aspects of combating the financing of extremist and terrorist activities.

The aim of the research is to improve strategic directions of the efforts to counter extremist and terrorist activities, to identify and characterize the ways of organizing financing of extremist and terrorist activities.

The existing situation in the field of countering extremist and terrorist activities is complicated by the high latency of these crimes, by ignoring the state borders, by the national and international legislation, by improving ways of conducting extremist and terrorist activities, etc.

The intensity of terrorist and extremist activities directly depends on the level of their financial, material and technical support. In this regard, the fight against the financing of terrorism and extremism is one of the priority tasks of the bodies of internal affairs, and the "freezing" of the assets of terrorist organizations and the closure of the channels for financing terrorist and extremist activities are recognized as the most important tools for combating international terrorism. However, until now the detectability and, accordingly, the disclosure of these crimes remain at a low level. It is even more difficult to prevent such actions at the stages of preparation or attempt to commit a crime.

2 The Legal Basis for Countering the Financing of Terrorist and Extremist Activities

Counteraction to the financing of extremism and terrorism is based on the provisions of the Constitution of the Russian Federation, the regulatory legal acts of the Russian Federation and a number of international legal acts. The legal basis for combating the financing of extremism and terrorism includes:

- The Criminal Code of the Russian Federation of June 13, 1996, No. 63-FL;
- Federal Law No. 114-FL of 25.07.2002 "On Counteracting Extremist Activity";
- Federal Law No. 35-FL of 06.03.2006 "On Counteracting Terrorism";
- Federal Law No. 115-FL of 07.08.2001 "On Counteracting the Legalization (Laundering) of Proceeds from Crime and Financing of Terrorism";

- Decisions and orders of the Government of the Russian Federation, as well as regulatory legal acts of the Central Bank of the Russian Federation and the
- Federal Service of Financial Monitoring.

In Russia, the responsibility for financing terrorist activities is provided for in the Article 205.1 of the Criminal Code of the Russian Federation "Assistance to Terrorist Activities" [2]. In accordance with note 1 to the Article 205.1 of the Criminal Code of the Russian Federation, financing of terrorism is understood as the provision or collection of funds or provision of financial services with the knowledge that they are intended to finance the organization, preparation or commission of at least one of the crimes provided for in the Articles 205, 205.1, 205.2, 205.3, 205.4, 205.5, 206, 208, 211, 220, 221, 277, 278, 279 and 360 of the Criminal Code of the Russian Federation, or for financing or other material support for a person to commit at least one of these crimes, or to provide an organized group, an illegal armed group, criminal community (criminal organization), established or being established to carry out at least one of these crimes.

In accordance with the clause 16 of the Resolution of the Plenary Session of the Supreme Court of the Russian Federation of February 9, 2012, No. 1, the financing of terrorism, along with the provision of financial services, should be recognized as the provision or collection of not only money (in cash or cashless form) but also material assets (for example, uniforms, equipment, means of communication) with the knowledge that they are intended to finance the organization, preparation or commission of at least one of the crimes provided for by the Articles 205, 205.1, 205.2, 206, 208, 211, 277, 278–279 and 360 of the Criminal Code of the Russian Federation, or to provide an organized group, an illegal armed group, a criminal community (criminal organization) established or being established to commit at least one of these crimes (for example, systematic payments or a one-time contribution to the general cashier, purchase of real estate or payment of its rent, provision of funds intended for bribing officials) [2].

Based on the definition given in the Article 205.1 of the Criminal Code of the Russian Federation, any means or financial services belonging to the terrorist financing entity may be used to finance terrorism.

Responsibility for financing extremist activities is provided for in the Article 282.3 of the Criminal Code of the Russian Federation "Financing Extremist Activity" (introduced by the Federal Law No. 179-FL of June 28, 2014) and consists in providing or collecting funds or providing financial services that are knowingly designed to finance the organization, preparation and commission of at least one of the extremism-related crimes, or to ensure the activities of an extremist community or an extremist organization.

3 Organization of Extremist and Terrorist Activities Financing

Ensuring the effectiveness of the law enforcement agencies in countering the financing of terrorism and extremism is impossible without a comprehensive study and a clear understanding of the organization of financing extremist and terrorist activities, including in a transnational aspect [1, 3, 4]. The establishment of methods, as well as sources and channels for financing terrorism and extremism, will facilitate the development of an algorithm of the law enforcement actions to identify and interdict them in order to bring the perpetrators to criminal liability.

The sources and channels for financing terrorism and extremism are economic entities (individuals, individual entrepreneurs and organizations, public associations that do not have the status of a legal entity), as well as their activities aimed at providing or collecting funds (money, tangible assets) or the provision of financial services with the knowledge that they are intended to finance terrorism and extremism.

In accordance with the grounds specified in the clause 6 of the Federal Law of 07.08.2001 No. 115-FL as of 2017 the "List of Organizations and Individuals for Whom There is Evidence of Their Involvement in Extremist Activities or Terrorism", published by the Federal Service of Financial Monitoring in its official website (http:.www.fedsfm.ru), comprises about 7 thousand individuals and legal entities, including: 7558 individuals and 88 organizations inserted in the national list; 411 individuals and 94 foreign organizations inserted in the international list. Since early 2016, credit institutions have taken measures to freeze the funds of these entities for more than 24 million rubles.

The empirical base subjected to the analysis comprises orders on the initiation of criminal cases, court rulings, interviews of the employees of the law enforcement agencies operating in this area in 16 regions of the Russian Federation. It has allowed to estimate the positive experience of the law enforcement agencies in countering extremist and terrorist activities and to identify the channels for financing these criminal activities.

3.1 Delivery of Funds Using the Facilities of Credit and Financial Institutions and Payment Systems

Banks transfer funds to bank accounts without opening bank accounts in accordance with the federal law and regulations of the Bank of Russia which have established forms of cashless payments made by payers, recipients of funds, as well as by persons or organizations with an authorized access to the bank accounts of payers.

Money transfer is carried out within the following forms of non-cash settlements: settlements by payment orders; payments under the letter of credit; settlements by collection orders; payments by checks; settlements in the form of transfer of funds at the request of the recipient of funds (direct debiting); calculations in the form of transfer of electronic funds.

To transfer and launder money, terrorists and extremists use such banking operations as the opening of all types of accounts; crediting and delivery of funds; cashing out; making payments; collection; attraction and placement of deposits; provision of loans; issuance and servicing of plastic cards; transfer of funds to correspondent banks; turnover of money through currency exchange points; opening, advising and execution of letters of credit; collection of checks and bills; connection to the "Tele-bank" system with additional protection; maintenance of accounts of front companies and controlled companies, including those registered abroad, primarily offshore.

The main task of the entities carrying out the financing of terrorism through banking structures and payment systems is to ensure that the conducted operation does not fall into the category of controlled ones [5].

Transferring funds for the financing of terrorism and extremism through banks are characterized by ensuring the anonymity of transaction participants and concealing the true purposes of their fulfillment. For this, a wide range of "tools" is used: "front" firms that do not conduct real financial and economic activities; a broad stratum of persons who lead an antisocial lifestyle and are ready to act as a leader and founder of a commercial organization for a small fee, to obtain a plastic card or a knowingly irrevocable loan using their personal data. The above situation is exacerbated by the existence of credit institutions that provide assistance for conducting doubtful financial transactions, as well as by the imperfection of existing control procedures that do not allow to fully trace the real objectives of financial transactions.

One of the most common forms of financing of terrorism and extremism in the Russian Federation is the functioning of organizations whose settlement accounts are used to transfer funds under the guise of paying non-existent goods and services or individuals who open accounts in their own name and then transfer cards to the terrorists and extremists, and if necessary supplement accounts for the financing of criminal activities.

The intensive development of new forms of financial transactions (e-wallets, electronic money, instant transfers), and the internationalization of financial transactions provide ample opportunities for uncontrolled remittances, including for terrorists and extremists. It should be noted that one of such channels—alternative payment systems—has a specificity. Despite the legitimate nature of their origin and functioning, the alternative payment systems are not subject to official control, and the facts of financing terrorism and extremism with their use can be identified only with the help of operational capabilities of law enforcement agencies.

The possibilities of depersonalization and unauthentic registration of virtual transfer systems make them the most popular channel for financing terrorism and significantly complicate the work of the law enforcement agencies in this area.

In this context, we cannot fail to mention such a channel as the conduct of financial transactions through the Internet [6, 7]. One of the most used channels for financing terrorism and extremism was the use of a new technology—the so-called crowdfunding, that is, a massive collection of voluntary donations using the details of accounts published on the Internet.

Such schemes make the financing of terrorist activities more accessible not only for terrorists but also for donors—radical citizens who were previously unable to provide assistance to terrorist organizations and are not associated with them.

3.2 Delivery of Funds Through Couriers, International Charitable and Other Non-profit Organizations

The extensive geography of the activities of charitable and other non-profit organizations creates ample opportunities for the transfer of funds for financing of terrorism and extremism. Representative offices of such organizations currently operate in almost all cities of Russia. In order to counteract the illegal activities of such organizations, the Ministry of Justice of Russia has developed a "register of non-profit organizations" performing the functions of a foreign agent", which includes 89 non-profit organizations, as of 2017.

3.3 Delivery of Funds Through Alternative Money Transfer Systems (AMT)

The most famous and widespread of such systems is "hawala" [8–10]. The word "hawala." of Arab origin means "trust" or "exchange". In South Asia, this model is known as "hundi"; in South America, as "kasa de kambyo" ("house of transactions"); the Thai version is "phu kuan" ("house of messages"); the English in China used "chitibanking"; and the Chinese themselves, who traded mainly rice and tea, developed their system of "flying money" – "fay Chien". These systems are currently successfully operating in most countries of the world, and their informal centers are London and Dubai.

The struggle with the AMT systems is complicated by the volume of rather small transactions within such systems, but their number is quite large—according to different estimates, they spend 200 to 500 billion dollars per year, in connection with which the alternative money transfer systems play a crucial role in the economy of the underdeveloped countries, servicing migrants' remittances to their families. In Pakistan alone, from 2 to 5 billion dollars are transferred through this system every year [8].

As a result of a joint study of the alternative money transfer systems, the International Monetary Fund and the International Bank for Reconstruction and Development have concluded that such systems meet existing needs by offering a competitive and efficient mechanism for transferring funds, though the anonymity of the systems makes them attractive for the use as a method of transfer of funds for financing of terrorism [9].

Cells of the alternative money transfer systems are present in many large cities of Russia, and their locations are situated in hotel complexes, in the territory of large shopping centers and markets, as well as near checkpoints across the state border of the Russian Federation.

3.4 Classification of Terrorism and Extremism Financing Sources

Any means or financial services belonging to a terrorism financing entity can be used to finance terrorism and extremism.

Sources of financing of terrorism and extremism can be classified according to the following grounds:

- depending on the subject of financing terrorism and extremism;
- depending on the nature of the activity, the result of which are funds for the financing of terrorism and extremism;
- depending on the type of funds and (or) financial services used to finance terrorism and extremism (cash and non-cash funds, securities, precious stones and metals, vehicles, buildings, facilities, other movable and immovable property, as well as property rights).

Thus, all that is used not only to finance the commission of specific terrorist and extremist acts but also what helps to sustain the life of terrorist and extremist organizations and (or) individual terrorists and extremists can be recognized as means for financing terrorism and extremism [10, 11].

So, the means of financing terrorism and extremism comprise both, the real estate objects (for example, a country house) provided for the living of terrorists, and the income (in whole or in part) received in cash or in kind of renting an apartment house to third parties, if such income or part of it is directed to the needs of a terrorist and extremist organization.

At the same time, cash and non-cash funds have been and continue to be the main type of funds used to finance terrorist and extremist organizations [12, 13].

The funds allocated for the financing of terrorism and extremism can result from both legal and illegal activities. Depending on their nature, the sources of financing of terrorism and extremism can be divided according to their origin:

- legal: profits taxes paid by entrepreneurs, donations of individuals, financing by interested organizations;
- illegal (criminal): illegal business, extortion, robbery, sale of property obtained by criminal means, sale of weapons and drugs, kidnapping, prostitution, etc.

Depending on the location of the financing entity and the location of the means of financing terrorism and extremism, the sources of financing are divided into external (the subjects and means of financing are located outside the Russian Federation) and internal (subjects and means of financing are located in the territory of the Russian

Federation), mixed (subject of financing is located in the territory of the Russian Federation and the funds are located abroad or vice versa).

Internal means of financing terrorism and extremism include:

1. funds received as a result of ordinary, economic and official crimes:

    a. theft, robbery, bribery and extortion (money from officials and businessmen);
    b. trade in arms, drugs and the organization of prostitution;
    c. manufacturing and marketing counterfeit banknotes and excise stamps, organizing illegal banking activities;

2. legal business under the control of members of illegal armed groups (income derived from legal activities in the budgetary sphere, the construction sector, agriculture, gold mining and turnover of precious metals, oil production and its processing in petroleum products);
3. help from relatives and friends (funds coming from the ethnic diasporas located in the most economically developed regions of the Russian Federation and other countries);
4. voluntary donations;
5. self-financing of militant terrorists, when people planning to travel outside the state to participate in terrorist activities, received knowingly irrevocable loans or sold their real estate.

External means of financing terrorism and extremism are:

1. foreign non-profit non-governmental organizations and non-profit organizations;
2. income from the shadow business controlled by the members of illegal armed groups located abroad, and crime-related funds of the transnational nature;
3. voluntary donations of Muslims from different countries is the most important source of financing for all Islamic groups (regular "zakat" and voluntary "sadaka" donations of believers for the charity needs). They are considered compulsory for all the faithful Muslims. Although "zakat" is paid by private and legal entities on a regular basis, it has traditionally not been legally regulated by the authorities (especially secular ones) and is not subject to tax, audit and any other verification by the state.

At the collection stage (in mosques, Islamic banks, etc.), these funds are absolutely legal. They are managed by various Islamic funds and charitable organizations, such as the International Islamic Relief Organization in Saudi Arabia or the World Muslim League, which has more than 100 branches in more than 30 countries. Institutions of the Islamic banking system also play an important role.

In accordance with the norms of Islam, regular donations in the form of "zakat' are officially intended for various forms of charity, as well as for propaganda of Islam, printing, and distribution of the Koran and construction of mosques, etc. In practice, some of these funds can be redistributed to the needs of Jihad and its leading radical Islamist groups. Thus, even those of the radical Islamic organizations that conduct terrorist activities and are outlawed are often financed from legal sources. Financing, in this case, is carried out according to the principle of not laundering dirty money, but polluting the clean money.

It is important for the law enforcement officers to identify what methods and ways are used to transfer the relevant assets from the source of financing of extremism and terrorism to the recipient, i.e. the terrorist or extremist organization or its individual participant, and to identify the channel and source of financing of extremist and terrorist activities. It is necessary for the prevention of crimes under the Articles 205.1, 282.3 of the Criminal Code of the Russian Federation.

4 Conclusion

Summarizing the above, it should be specially noted that the most important direction in countering terrorist and extremist activities is neutralizing the ways of financing them and eliminating the sources of criminal proceeds. In modern conditions, the activities of law enforcement agencies should be aimed at optimizing the forces and means of countering terrorist, extremist activities [14, 15], while building on an innovative interdisciplinary approach and focusing on improving criminal justice in a transnational context [16–18]. In this regard, the attention was brought to the need for a comprehensive research and implementation of a number of significant trends in the activities of the law enforcement agencies:

1. Optimization of the mechanism of legal regulation of counteraction to terrorist and extremist activity.
2. Information support for countering terrorist and extremist activities.
3. Organization of international cooperation in the field of countering terrorist and extremist activities.
4. Optimization of preventive activities in the context of countering terrorist and extremist activities.
5. Resource support for countering terrorist and extremist activities.

References

1. Decree of the President of the Russian Federation of December 31, 2015 No. 683 "On the National Security Strategy of the Russian Federation". Legislation Acts of the Russian Federation. 2016, No. 1 (Part II). Art. 212
2. The Criminal Code of the Russian Federation of June 13, 1996, No. 63-FL. Legislation Acts of the Russian Federation. 1996, No. 25, Art. 2954
3. Federal Law "On Counteracting Terrorism" of March 6, 2006, No. 35-FL. Legislation Acts of the Russian Federation. 2006, No. 11, Art. 1146
4. Federal Law "On Countering Extremist Activity" of July 25, 2002, No. 114-FL. Legislation Acts of the Russian Federation. 2002, No. 30, Art. 3031
5. Federal Law "On Combating the Legalization (Laundering) of Proceeds From Crime and Financing of Terrorism". Legislation Acts of the Russian Federation. 2001, No. 33 (Part I). Art. 3418

6. Janken, K.: Assessment of reliability of sources of the global Internet. Polizei-heute. No. 4 (2010)
7. Stepanova, E.K.: Counteraction to the financing of terrorism. International Processes. No 2(2011)
8. Khokhlov, I.I.: The shadow and light of the hawala. Financial system of the Middle Ages in the context of globalization. Military Industrial Courier. No. 16 (182). (2007)
9. Kuznetsova, O.S.: Development of institutions to counter the economic foundations of international terrorism: thesis for the degree in economic sciences. M., pp 53–60 (2006)
10. Algiers Memorandum on Good Practices on Preventing and Denying the Benefits of Kidnapping for Ransom by Terrorists. https://www.thegctf.org/documents/10162/159874/Algiers+Memorandum Englis.pdf. Accessed 7 July 2018
11. Coleman, T., Bartoli, A.: Addressing Extremism. The International Center for Cooperation and Conflict Resolution (ICCCR), Teachers College, Columbia University; The Institute for Conflict Analysis and Resolution (ICAR), George Mason University, (2012)
12. Lenain, P., Donturi, M., Koen V.: The Economic Consequences of Terrorism. OECD Economics Department Working Papers, No. 334, OECD Publishing, Paris. http://dx.doi.org/10.1787/51 1778841283. (2002)
13. Müller Sebastian, R.: Hawala: An Informal Payment System and Its Use to Finance Terrorism. Saarbücken, VDM Verlag Dr. Müller, p. 83 (2006)
14. Cramer, D., Brown, A.A., Hu G.: Predicting 911 Calls Using Spatial Analysis. In: Lee R. (eds) Software Engineering Research, Management and Applications 2011. Studies in Computational Intelligence, vol 377. Springer, Berlin, Heidelberg (2011)
15. Wikström, P.O.H.: Routine Activities, Area Crime, and Victimization Patterns. In: Urban Crime, Criminals, and Victims. Research in Criminology, pp 185–236. Springer, New York, NY, (1991)
16. International Convention for the Suppression of the Financing of Terrorism (adopted by resolution 54/109 of the General Assembly of the United Nations on 9 December 1999)
17. The Shanghai Convention on Combating Terrorism, Separatism and Extremism (Shanghai, 15 June 2001)
18. The Recommendations of the Group of Development of Financial Measures of Struggle Against Money-Laundering (FATF)

Radio-Electronic Warfare as a Conflict Interaction in the Information Space

Alexander Kupriyanov, Anatoly Ovchinsky, Alexander Betskov and Vadim Korobko

Abstract Contemporary conflicts between countries are increasingly taking place not in the direct armed clashes, but in the information field. In this connection, the analysis of the possibilities of the application of the radio-electronic means during the information conflict is of scientific interest. The possibilities of applying modern radio-electronic aids in the course of information conflicts are considered and analyzed. Technological base of modern information communication systems is radio electronics, therefore technically such conflictual interaction is provided by the methods and aids of electronic warfare (EW). The capacious synthetic concept of EW comprises radio-electronic reconnaissance (ER), which identifies the enemy's EA and extracts information about them necessary for the ES, as well as radio electronic masking (EM), opposing the radio electronic reconnaissance of the enemy in the information conflict. The conceptual bases and prospects of the development of the EW subject area are discussed on the basis of modern technological achievements, digital methods of signal processing in information communication systems, management of information systems and media. The peculiarities and directions of the development of conflict interaction in the information space are considered.

A. Kupriyanov (✉)
Moscow Aviation University (National Research University), Moscow 125993,
Russian Federation
e-mail: aik@mai.ru

A. Ovchinsky
Moscow University of the Ministry of Internal Affairs
named After V. Ya. Kikot, Moscow 17437, Russian Federation
e-mail: sunobor@yandex.ru

A. Betskov
Academy of Management of the Ministry of Internal Affairs of Russia,
Moscow 125993, Russian Federation
e-mail: amvd-6@bk.ru

V. Korobko
All-Russian Public Organization «Business Russia»
Moscow, Moscow, Russian Federation
e-mail: vkorobko@mail.ru

© Springer Nature Switzerland AG 2019
A. G. Kravets (ed.), *Big Data-driven World: Legislation Issues and Control Technologies*, Studies in Systems, Decision and Control 181,
https://doi.org/10.1007/978-3-030-01358-5_16

Keywords Electronic warfare · Electronic masking · Electronic reconnaissance Electronic suppression

1 Introduction

According to the definition of the various directories and encyclopedic dictionaries [1], the "electronic warfare" (EW) is, either a kind of an armed struggle or a kind of operational (combat) support for the actions of the armed forces. But, as often happens, the definition of the term does not keep pace with the changes in the essence of the subject area, using the objects designated by this term. So, for the present time, the definition of the term EW is clearly outdated. First, it is well known that EW aids are used in pure peacetime. Secondly, the methods and aids of EW are used not only by the armed forces. Therefore, according to the modern views, the term "electronic warfare" refers to the conflict interaction of radio electronic systems in the information space. Radio electronic systems, which are the basis of the information technologies, define the main features of the world civilization in the twenty-first century.

It is the conflictual interaction of radio electronic systems and aids that should be considered as a driving force for the development and improvement of the info media. The most traditional and obvious are the manifestations of the conflict of information systems with nature. Nature limits the noise immunity of information transmission via communication channels, data transmission speed, the efficiency of information and measurement systems, the performance of information sources, the capacity of data banks and other information resource stores.

But nature, as A. Einstein has asserted, is "sophisticated, but not malicious." To date, constructive and convenient models of the influence of natural factors on information systems have been developed and introduced into the practice of creating and operating these systems. Such models, which form the basic content of the theory of information transfer, considerations about the potential noise immunity, and other conceptual provisions, are widely used by the creators and users of information communication systems and information technologies. Therefore, engineers are now winning in the competition of the collective engineering intelligence of the creators of information systems with nature. The conflict of information systems with counteracting systems of anthropogenic origin is another matter. This is the interaction of the conflicting interests of different groups and coalitions of participants in information processes. The outcome of this conflict is not at all as unambiguous as in conflictual interaction with nature. This conflict interaction in the information space, the character of which almost always acquires antagonistic features, nowadays should be called the electronic warfare [2–4].

2   The Structure of Electronic Warfare

The structure of the complex synthetic concept of "electronic warfare" can be illustrated by the presented graph (Fig. 1).

This graph shows the structuring of the subject area, denoted by the EW term, as a set of activities interconnected by purpose, tasks, place and time, actions aimed at identifying the enemy's electronic aids and systems (EA) in the information conflict (EW), their radio electronic suppression (ES), as well as radio-electronic masking (EM) and protection (EP) of their radio electronic systems and aids against the enemy's negative and destructive actions.

This dialectical unity and conflict interaction of such opposites as the ER and EM, the ES and the EP, basically determine the dynamics of the intensive development of aids and methods of electronic warfare. It will not be a big exaggeration to say that the unity and struggle of these opposites largely determine the nature of the current stage in the development of the theory and technology of information systems and information technologies. Moreover, it changes our perception of the significance of electronic warfare for different spheres of life of the modern postindustrial, information stage of the development of civilization.

As already mentioned above, according to modern views, electronic warfare is the result of the transformation of combat operations into an independent kind of activity that has spread beyond the boundaries of a purely military sphere. Methods and aids of EW are based on the information systems which function in the conditions of conflict confrontation and provide state, regional, corporate and military control, and operation of economic infrastructure.

The functioning of modern information systems of all known types, such as transmission and retrieval of information, radio control and destruction of information, involves the use of all known physical fields (electromagnetic, acoustic, seismic, etc.) that are known to date. It is supported by the use of oscillations in an extremely wide range - from super long radio waves and infra-low frequency oscillations of the earth's crust to ultraviolet and gamma radiation waves. All these fields are informative for technical reconnaissance aids [5].

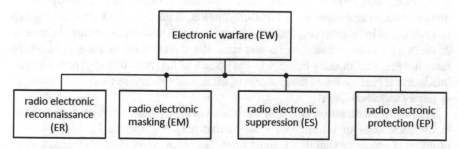

Fig. 1   Structure of "electronic warfare" concept

Requirements to ensure the effectiveness of obtaining information by the technical aids of ER provide for the use of various platforms for the placement of these aids. In addition to the traditional stationary, mobile and portable ER aids, aeronautical aids (including on-board drone facilities) are widely used. Radio electronic reconnaissance makes use of and constantly raises the efficiency of space-based facilities of delivering information to the consumer. The outer space is extraterritorial, and the use of the space reconnaissance aids is not constrained by state borders. Also, the possibilities of reconnaissance to use space platforms for the deployment of technical aids are not affected by the restrictions imposed by the terrain in the field of the intelligence interests.

The problem of EW is very clearly stratified. It stipulates the division into levels and the unification of multiple levels of description. These levels represent the physical and technical principles of the construction and functioning of the aids involved in the information conflict. The descriptions and models of the system design principles and organization of interaction of these aids are fixed; principles of tactics and operational art of using EW aids and methods in peacetime and at different stages of the development of armed conflicts are determined.

The manifestations of the conflict of information systems and information destruction systems are very diverse. They combine simple interactions, like interference and interference protection, as well as sophisticated ways to ensure the reliability and authenticity of information resources in the modern environment of digital information interaction.

3 Information Destruction as the Main Aim of EW

The destruction of information is almost the oldest, but also the current direction of EW aimed at the electronic counteraction or defeat. It can be achieved not only by the interference [6]. The EW arsenal provides methods of technical disinformation. These methods are used to create distorted representations about the information media and provide for a controlled demonstration of false unmasking characteristics of electronic aids, objects, and conditions, deliberate intrusion into the enemy's information systems and networks for misinformation, distortion of information, signals and call signs. In other words, they are applied to create false representations (images) of one's own information systems and aids. But the creation of images, including false images, traditionally belongs to the sphere of art. And this idea convincingly indicates the emergence of the problem of electronic warfare beyond the boundaries of purely technical areas.

An important direction of EW is a decrease in the visibility of signals circulating in the backbones of information systems and networks for the ER aids [4]. The problem of reducing visibility is most important when applied to radar but is by no means limited to the scope of this subject area. The solution of the complex problem of reducing the visibility of the signals of EM is far from complete and involves conducting research in the fields of electrodynamics, aerodynamics, hydrodynamics, reliability, and technology of production of information media and their carriers. The search for new principles for the construction and organization of the work of information sensors which ensure the secretiveness of the functioning of control systems is equally relevant.

The aids and systems of the EW function in close interaction with information media of a different purpose. For example, a phased array and radar transmitters can be used for an EW at the stage of penetration of antiaircraft and anti-missile defense lines. And when an impact is struck, the transmitters of the active jamming station can be used to build up the radar's energy potential or to solve the weapons control tasks. In the integrated radio electronic systems, the resources of various systems will be redistributed among themselves to achieve a synergistic effect and increase the efficiency of solving specific targets. Of course, such integration of EW assets with other systems and aids in the information space provides for the technological unity of the foundations of their construction. Therefore, digital methods for the formation and processing of signals, the formation of influences which control the operation of information systems and aids are widely and almost exclusively used in the construction of modern and promising aids of electronic warfare.

Usually, the methods and means of electronic countermeasures under consideration envisage working with low or high energy. The electronic countermeasures of the radio electronic suppression work with low energy. Of course, the interference energy is small only in the sense that its levels are commensurable with the energy of the suppressed signals. Such EW aids carry out information suppression and disinformation impact. The physical defeat of the enemy's radio electronic and, in general, information aids (for example, the defeat by high-precision weapons, which are targeted at the sources of radio emission accompanying the operation of information systems) involves scattering of much higher energy. But the latest technological advances in the development of high-current electronics allow the creation of electronic warfare equipment, occupying an intermediate energy space between information and physical, fire destruction systems. These are aids of functional defeat. They produce destructive effects on the radio electronic components of enemy information systems by strong electromagnetic fields. Aids of functional damage are used to generate powerful electromagnetic pulses devices of relativistic electronics and electromagnetic ammunition equipped with explosive magnetic generators or shockwave source of electromagnetic radiation. In such sources, the chemical energy of a condensed explosive of ammunition and the electrical energy stored in superconducting systems are transformed into powerful electromagnetic radiation capable of disabling radio electronic aids in a radius of tens or even hundreds of meters.

4 EW Effect on Computer Systems

The widespread introduction of digital technology into information communication systems [7], the operation of which is supported by computer systems and aids, has stimulated the emergence of a new type of electronic suppression based on the use of destructive program influences.

Firstly, these are computer viruses (that are not themselves programs), which are embedded in programs and disrupt the operation of computer systems, distort data. Computer viruses, as a rule, have the ability to reproduce and multiply themselves, and this property poses a real threat to information and complicates the detection of information attacks and prevention of their negative consequences. The events of recent years have shown how deliberate use of computer viruses can disrupt the functioning of the systems operating under computer control. This is especially important for the protection of the objects of vital concern and information infrastructure [8].

Secondly, there are such EW aids as destructive software effects on digital computing systems, otherwise known as program bookmarks. Such malicious programs differ from the computer viruses by operation principle. They do not have the ability to reproduce or to self-duplicate themselves. But, getting unauthorized access to data in memory or in the exchange channels, they intercept and/or distort this data.

Aids of electronic suppression and electronic masking are created and used to reduce the risks of information loss and improve the efficiency of information systems. Therefore, the aids and methods of the of electronic suppression and masking appropriate for EW should be considered as the most important components of the armed struggle as well as of the protection of the state information security [9].

High-tech aids, using both high-precision and information weapons, largely affect the change in the nature of the armed struggle. The experience of the recent local wars demonstrates that in any operations, high-precision weapons are primarily used to disrupt the functioning of the information infrastructure components of the state and military control complexes. Thus, in the wars in Iraq in 1991, 1998 and 2003, in Serbia and Kosovo in 1999 from 60 to 90% of the enemy's objects of information infrastructure were defeated with the help of high-precision weapons. Therefore, based on the analysis of the experience of conducting EW in local wars and armed conflicts of recent times, one can justify the assumption that the military victory will be reached only by the side that is better prepared for action in the context of information conflict and, in particular, in the context of electronic warfare. To win a victory, any country needs to have more effective aids for radio electronic suppression and reconnaissance as well as better operational facilities. It is also necessary to have better trained competent specialists who are able to operate in a conflict radio electronic environment.

5 Conclusion

We can also make a conclusion that the electronic warfare has obviously transformed from episodic actions to create radio interference to individual radio links into a set of theoretically developed and practically implemented techniques for developing a rational strategy of behavior in an information conflict. It has changed from the use of individual aids of staging radio interference to the complex application of information countermeasures, systems of operational, tactical and strategic radio-electronic reconnaissance [10].

The format of a brief chapter does not allow us to consider the complex problem of electronic warfare with a more thorough degree of detail. Nevertheless, the authors hope that they have managed to list the main circumstances that determine the relevance of this problem, its practical significance for the development of strengthening Russia's economic and defense power, the reliability and security of the Russian national information structure.

References

1. Collins English Dictionary—Complete and Unabridged, 12th Edition 2014 © HarperCollins Publishers 1991, 1994, 1998, 2000, 2003, 2006, 2007, 2009, 2011, 2014
2. Kupriyanov, A.I., Shustov, L.N.: Electronic Warfare. Fundamentals of the Theory. Moscow, Vuzovskaya book (2017)
3. Davison, N.: Directed energy weapons. In: 'Non-Lethal' Weapons. Global Issues Series, pp. 143–185. Palgrave Macmillan, London (2009)
4. Lambrechts, W., Sinha, S.: Electronic countermeasures, and directed energy weapons: innovative optoelectronics versus brute force. In: SiGe-based Re-engineering of Electronic Warfare Subsystems. Signals and Communication Technology, pp. 133–166. Springer, Cham (2017)
5. Quisquater, J., Samyde, D.: Radio frequency attacks. Encyclopedia of Cryptography and Security, pp. 1015–1021. Springer, US (2011)
6. Gandolfi, K., Mourtel, C., Olivier, F.: Electromagnetic attacks: concrete results. In: Koc C.K., Naccache D., Paar C. (eds.) Proceedings of the Cryptographic Hardware and Embedded Systems (CHES), Paris. Lecture Notes in Computer Science, vol. 2162, pp. 251–261. Springer (2001)
7. Kravets, A.G., Bui, N.D., Al-Ashval, M.: Mobile security solution for enterprise network. Commun. Comput. Inf. Sci. **466** CCIS, 371–382 (2014)
8. Mishra A.K., Verster R.S.: Compressive Sensing Based Algorithms for Electronic Defense. Signals and Communication Technology, p. 184. Springer International Publishing (2017)
9. Lambrechts W., Sinha S.: SiGe-based Re-engineering of Electronic Warfare Subsystems. Signals and Communication Technology, p. 329. Springer International Publishing (2017)
10. Paliy, A.I., Kupriyanov, A.I.: Essays on the History of Electronic Warfare. The University Book, Moscow (2006)

Part IV
Practical Aspects and Case-Studies of Legislation Regulation and Control Technologies Development

Mechanisms for Ensuring Road Safety: The Russian Federation Case-Study

Viktor Kondratiev, Alexander Shchepkin and Valery Irikov

Abstract Traffic accidents cause the Russian economy enormous damage, which in recent years is 2.2–2.6% of the country's gross domestic product. Such damage is comparable with the contribution to the gross domestic product of the individual sectors of the national economy. The scale and nature of the road safety problem in the country, its social, economic and demographic consequences have a significant impact on the national security of the country, and the task of ensuring road safety represents an independent state problem. The methodological basis of the research is the general management theory, elements of the theory of decision-making and risk management, the theory of program-target management. The chapter considers the problems of control in the field of road safety. A description of mathematical models and mechanisms for ensuring road safety is provided, including the mechanisms for the integrated assessment of the activities of the Traffic Control Department, methods for developing programs to improve the level of road safety with due regard to the reliability factor. The road safety system functions within the framework of the national economic complex of the country and in the society, being a socio-technical system of the society and the state. Therefore, it is considered that the road safety system, like any other similar system, cannot exist without an appropriate set of supporting functions, such as medical, financial, economic, technical, personnel, regulatory, informational, supervisory, scientific, social–political, special.

V. Kondratiev (✉)
Moscow State Automobile and Road Technical University, Moscow 125319,
Russian Federation
e-mail: k-051310@mail.ru

A. Shchepkin · V. Irikov
Institute of Control Science, Russian Academy of Sciences named after V.A.Trapeznikov,
Moscow 117342, Russian Federation
e-mail: sch@ipu.ru

V. Irikov
e-mail: irikov41@mail.ru

© Springer Nature Switzerland AG 2019
A. G. Kravets (ed.), *Big Data-driven World: Legislation Issues and Control Technologies*, Studies in Systems, Decision and Control 181,
https://doi.org/10.1007/978-3-030-01358-5_17

Keywords Road safety · System approach · Federal target program
Control mechanism · Program-target planning · Structural-functional model
Integrated assessment

1 Introduction

1.3 million people are killed, and 50 million people are injured in road traffic accidents (RTA) in the world for one year. Economic damage from road accidents is more than 500 billion US dollars.

According to the forecasts of the World Health Organization (WHO) by 2030, road traffic traumatism (RTT), if not taken the appropriate measures, can become the fifth among the main causes of death [1].

In the Russian Federation, the level of traffic accidents remains high. Over the past 10 years, more than 300 thousand people died in an accident, which is equivalent to the population of an average regional center. Practically 1/3 of the dead are people of the most active working age.

Compared with European countries, the accident rate in the Russian Federation is characterized by the highest risk of death, high severity of consequences.

Road traffic crashes (RTC) cause damage to the Russian economy, which is about 2.0% of the country's gross domestic product (GDP). Such damage is comparable with the contribution to GDP of the individual sectors of the national economy.

The scale and nature of the road traffic safety (RTS) problem, its social, economic and demographic consequences have a significant impact on the national security of the country; and the task of providing road traffic safety represents an independent state problem.

During the period of socio-economic reforms, the democratization of all aspects of public life, the change in the main accident rates was uneven.

The situation, similar to the Russian one, was observed in the countries of Europe in the 1970s of the last century. At present, in most countries with high and medium level of motorization there is a tendency of a consistent decrease in the basic indicators of RTT, which is the result of a long-term joint work of the subjects of management at all levels, Analysis of the state and structure of RTC in the country allows us to conclude that the main reason for the high rate of RTT is the low level of legal awareness and transport culture of road users.

The low discipline of drivers of motor transport is directly related to the failure of drivers to comply with the requirements of the current rules and regulations; mistakes in choosing the traffic regime because of low discipline or inadequate assessment of the development of the traffic situation.

This characteristic of drivers acquires special significance in the conditions of the unsatisfactory state of the road and transport infrastructure, constant growth in the number of vehicles, the complication of traffic conditions etc.

The decisive influence on the level of accident rate is rendered by drivers of vehicles belonging to individuals. Due to violations of traffic rules, more than two-thirds of the total number of accidents have been committed.

Within one year, the traffic police officers detect more than 60 million traffic offenses. At the same time, more than 500 thousand drivers are identified for driving in a state of intoxication.

In relation to the population size, the density of hard-surface roads in the country is about 5.3 km per 1000 inhabitants, which is significantly lower than in many economically developed countries. In the USA this indicator is equal to 13 km, in Finland—10 km, in France—15 km.

The basis (92%) of the domestic network of public roads is made up of territorial roads, the greater part (about 60%) of which belongs to one of the lowest—IV technical category by their parameters.

For example, the operational status of more than 60% of the length of regional, intermunicipal and local roads does not meet regulatory requirements.

The ongoing activities in the field of organization of vehicle and pedestrian traffic, with a few exceptions, are in many ways local, poorly interrelated and do not constitute a single citywide scheme.

In the country, there is a marked disproportion between the pace of development of the road network and the growth in the number of vehicles. In 2014, there were about 300 automobiles per 1000 inhabitants.

A negative characteristic of the domestic transport is a significant share of vehicles, having a long service life, including outside the established motor resource. So, there is a problem of "aging" of the domestic motor transport fleet. In 2014 the share of cars with a service life of over 10 years was 46.0%, trucks—61.9%, buses—44.9% and motor transport—90.4%, and the total share of "old" motor vehicles amounted to 51.2%.

The situation is aggravated by the fact that the level of active and passive safety systems does not meet modern requirements for domestic cars, especially those with a long service life [2].

Currently, more than 90% of the entire vehicle fleet is owned by individuals, which makes the possibility of carrying out preventive work extremely limited. The number of accidents and the number of injured due to the technical failure of the vehicle for the past 4 years have more than doubled. For this reason, the number of victims is growing for the fifth consecutive year.

According to the conclusion of the experts of the countries of developed motorization, drivers and passengers moving in old-fashioned cars are three times more exposed to the risk of being involved in an accident. In these conditions, the requirements for the efficiency of the vehicle safety system are significantly increased.

Along with the growth in the number of educational institutions for training of drivers in recent years (the number of which is approximately 8500), the existing system of these organizations has several shortcomings, the main ones of which are: departmental disunity, weak material and technical base, insufficient qualification of teaching staff, failure to fully implement the training programs and their imperfection.

2 Problems of Russian Policy in the Field of Road Safety

To date, Russia is only in the process of shaping the state policy in the field of road safety, which is a consequence of the absence of targeted long-term strategies for road safety in the previous years. A clear division and consolidation of the respective powers and responsibilities between the subjects of management at all levels (federal, regional and local) have not yet been achieved. The responsibility of state authorities and local self-government for the state of road safety is neither legally nor organizationally defined; as a result, there is a disunity and extremely inadequate coordination in the activities of government bodies. In many cases, all practical activities for the provision of road safety are limited only to the work of the state Traffic Control Department.

These circumstances lead to the fact that the management is not aimed at the achievement of goals. An untimely feedback system results in an untimely response of the control system to the changes in both external and internal environments.

The serious shortcomings in the organization of management determine ineffective financing of activities in this area. These shortcomings should, first of all, be attributed to the lack of clearly defined sources of funding at the federal, regional and local levels.

With the adoption of the Federal Targeted Program (FTP) for the period of 2006–2012s, the financing situation has improved. However, there is no holistic solution to the problem of creating a financing system.

Thus, during the implementation of the activities of this FTP, the adjustment of funding was carried out almost every year. As a result, the funds were reduced by the federal budget by 10%, by regional budgets—by 22.1%, and the total amount of financing was decreased by 15%. And the adoption of a new FTP to improve road safety for 2013–2020s, unfortunately, did not change the situation.

There are also significant shortcomings in the legal support of the organization of training and retraining of vehicle drivers. Issuance of licenses to legal entities and individuals for the training and retraining of vehicle drivers is carried out by the Ministry of Education and Science of Russia in the absence of both current and subsequent control over compliance with licensing requirements. At the same time, about two million new drivers annually complete the relevant training courses.

The forms of state and public influence on the traffic participants, aimed at the formation of socially significant stereotypes of the transport culture and the lawful conduct, have not been legally established. The situation is exacerbated by the generally low level of legal awareness, the lack of adequate understanding of the causes of the accident by the traffic participants, insufficient involvement of the population in the prevention of traffic accidents.

At the same time, the sanctions envisaged by the Code of Administrative Offences of the Russian Federation are mainly used for violations of traffic rules. The legislation does not practically use incentive measures as measures that stimulate drivers and pedestrians to comply with the traffic rules and the requirements of other regulatory documents on road safety issues.

However, the countries of developed motorization apply a very wide set of stimulating statutory measures for the encouragement of drivers and pedestrians (incentives through insurance instruments, "writing off" of penalty points, reduction in the period of deprivation of rights, etc.).

The FL "On Road Safety" defines "a program-targeted approach to activities to ensure road safety" as one of the basic principles of road safety.

The first FTP was the program "Improving Road Safety in Russia in 1996–1998s." The actual period for the implementation of the program was until 2001. In 2002, the subprogram "Road Safety" of the Federal Target Program "Modernization of the Russian Transport System" was approved, in which a broad range of measures to improve road safety was defined. At the same time, the result of the departmental organizational activities related to these programs, indicates their apparent setbacks, as, instead of the planned reduction in the number of deaths in 2002 by 895 people, the number of wounded by 5370 people, there was the growth of them recorded according to the results of the year. A similar situation of a significant increase in the main indicators of RTT, instead of their decrease, occurred in 2003 and 2004.

As a result, the advisability of program planning in the field of road safety was largely questionable. At the meeting of the government, it was noted that the subprogram "Road Safety" does not have the necessary impact on the state of road safety.

During the subsequent work, the analysis of the experience of program approaches to the organization of activities in the field of road safety has been conducted. It was stated that the program-target approach is an effective mechanism to address the problems of reducing RTT. But this approach requires some adjustments in the cause of development and implementation of the Federal Program. All this was taken into account when preparing the road safety program for 2006–2012s, and also in the preparation of the FTP for 2013–2020s.

At the present stage of the socioeconomic development of the Russian Federation, it is unreasonable (in the coming years) to set as a priority task of the achievement of the traffic accident rate corresponding to the level of accidents in the United States or other industrialized countries such as Great Britain, Sweden, and the Netherlands. In the first approximation, this can be argued by the following considerations. The dynamics of changes in indicators characterizing the level of RTT, to a certain extent, corresponds to the dynamics of changes in indicators characterizing the scale of other negative phenomena of public life, including criminal offenses, for example, such as suicide or premeditated murder.

The close connection in the change of such events with the changes in the accident rate is indicated in [3], where data are given about the general pattern of change in the number of suicides and the number of fatalities in an accident with a probability of 89%, the number of intentional murders and the number of fatalities in an accident with a probability of 95%.

Thus, the existing correlation (according to the law of large numbers) indicates the existence of a general trend in the dynamics of changes in indicators characterizing such different negative manifestations of social life, as accidents or killings (suicides). It is obvious that the instability of the conditions of social life, economic and political factors, the presence of crisis phenomena in the society affect the functioning of all systems of the state that have social subsystems. The traffic system, in which virtually the entire population of the country is involved, in this sense, cannot be an exception.

At present, unlike the industrialized countries of the West, the Russian Federation has the "other" level of the RTC, as it has not yet overcome the crisis phenomena in the socio-economic life of the state. Nevertheless, it is expedient to set the task of stabilizing the situation with traffic accidents, and then gradually reducing the level of RTT, that is, possibly reducing the danger of road traffic traumatism.

It is of some interest to analyze the main ideas in the field of road safety, that took place during the last 35 years in the Russian Federation, and earlier in the USSR, in fact from the time when this problem began to be designated as having great social and economic significance.

Previously, in the Russian Federation (before the collapse of the Soviet Union), mechanisms based on the concepts of a centralized system of road safety were in place.

At the highest level, various decisions were taken to improve road safety (often without an assessment of resource support). Further management—legally formalized in the form of interdepartmental documents—was carried out through ministries, departments and local bodies. Fundamental documents—decisions of the Central Committee of the CPSU and the Council of Ministers of the USSR (more often the resolution of the Council of Ministers of the USSR)—were adopted on average every five years.

At the same time, the concept of the content of preventive measures was based on the strategy of deterring the level of RTC. The essence of the strategy was that the rate of growth in the number of dead and wounded in an accident should have been less than the growth rate of motorization. In the quantitative sense, this meant that the relative accident rates, that is, the figures usually attributed to 10 thousand vehicles, were constantly decreasing.

The core of that strategy was based on the ideological idea of the absence of contradictions in socialist systems. Therefore, if we proceed from this logic, the accidents are of a purely technical nature and, therefore, transitory and temporary. At the same time, the scale of the road safety problem was presented as having no important state significance. In national economic planning, the state budget did not provide for a holistic solution to this problem.

Nevertheless, despite rather high rates of motorization in conditions of rigid centralization, disciplinary and party responsibility up to 1986–1988s, the state has managed to restrain and, in some years, to reduce the traffic accident rate. This can be illustrated by the results of the investigations carried out during 1972 and 1986 anti-alcohol campaigns. It is important to note that it was the fight against traffic accidents, and not against their consequences, i.e. the number of dead and injured in them.

At the same time, the effectiveness of this mechanism in the existing conditions of that time was quite high, which was reflected in the statistics of the RTA that has an annual 3–5% increase at a 10–12% increase in the auto fleet. In 1989, this scheme began to collapse due to socio-political reasons. As a result, during 1989–1991s, the traffic accident rate has dramatically increased.

Thus, the administrative "pressure" mechanism is able to exert only a limited influence on the rate of traffic accidents. This fact eventually has led to an actual confrontation between the traffic participants and traffic police officials.

The most noticeable in the transition period is the study [4], which has noted that the country has a high level of RTT. Eliminating the causes of RTC requires huge investments, which the country will be unable to do in the coming years. The authors of the research have considered the problem of "maximizing the level of road safety with limited financial resources". Their conclusion is formulated as the need to create an optimal system for road safety.

3 Structure of Road Safety Providing

Studies [4–7] are of the utmost interest. Their authors emphasize the "exceptional" complexity of the road safety area as an object of improvement, noting:

- the organizational and structural complexity associated with a lot of interacting entities, including those which are hierarchically coordinated in difficult-to-control conditions;
- variety of functional structures of different nature: vehicle, human-vehicle, complex, uniting large teams and complex technical means, information, etc.;
- multifunctionality, the complexity of intrasystem connections, the variety of local and social goals;
- presence of constantly remaining structural, normative, informational, methodological and other uncertainties;
- the stochastic character of the functioning processes associated with random deviations of the parameters of functional elements (equipment failures, personnel errors, random external influences), etc.

Highlighting a significant number of subsystems in the sphere of providing road safety (scientific, methodological, material, technical, technological, information, propaganda, etc.), the authors [5–7] define the organizational structure as a fundamental one. It is stated that the organizational support should establish such a structure of the traffic security management that could ensure the effective performance of functions by all the structures involved. The authors identify the control and supervisory subsystem as the most important element of the organizational support, which is the instrument for introducing the entire system into a controlled mode of operation.

To develop a long-term macroscopic forecast of the development of the traffic accident situation, an adaptive evolutionary model [8] is proposed, the essence of which is to combine the adaptation of traffic safety and the growth of the transport load. In this work, it is noted that mathematical models are rarely used in social, economic systems, as well as in psychology. The authors consider that reducing mortality in industrialized countries is a result of teaching the nation in the field of road safety.

It can be concluded that the authorities, scientists, and experts involved in solving problems of the road safety adhere to a certain stereotype in assessing the true causes of a high level of RTT.

According to many administrators and experts, the unsatisfactory state of the street-road network, its small extent, the poor quality of domestic vehicles, and the low discipline of drivers are the most common factors, which determine the high level of RTT.

Undoubtedly, these factors have a significant impact. However, analyzing the scale of the problem, the digital data characterizing the state of this problem, we are convinced of the need to reveal the deeper causes and factors that generate a high level of RTT.

A convincing argument confirming the importance of considering the social nature of the road safety problem is given in [4] when developing the forecast of the number of accidents. It is stated that "predictive models of accidents are models of a special type, in which the main controlling factor should be taken into account", that is, "the will of the relevant actors, the activity, and purposefulness of their will manifestation". It is also noted here that American experts received discrepancies with the actual development of events by a factor of 1.5 since they did not consider such social factors as "a qualitatively new attitude to road accidents, both on the part of the population and public services." They failed in developing a predictive model for determining the number of accidents, even though their calculations were based on "fine programs, detailed and reliable statistics made with the help of the computer of the last generation". The analysis of the traffic accidents shows that there are about several hundred causes which can be classified as the basic ones. At the same time, there are about several thousand additional conditions and circumstances contributing to the occurrence of accidents [9, 10]. This determines the complex nature of the problem, and the need to coordinate the efforts of the authorities of all levels, local governments, public associations and individuals in the prevention of road traffic accidents and their consequences

Thus, the complex approach to studying the problem of providing road safety is extremely important.

The emphasis on the fact that the high level of accidents in the country is largely determined by bad drivers, unfortunately, is not very convincing, since drivers are representatives of the country's population and members of its society. Rather, it should be said that economic instability and the worsening of the moral state of the society contribute to a drop in the transport discipline of the traffic participants.

From what has been said, the following conclusion can be drawn: the road transport complex of the country, which forms the road traffic system, functions inefficiently. Inadequate functioning of the traffic system leads to a large expenditure of resources per unit of profit received and, as a result, a high level of RTT, and, in general, to a high level of transportation costs in the organization and implementation of the transport process, transportation of people and goods. It follows that the current situation with accidents is largely determined by the generally poor performance of the road and transport complex management system, the shortcomings of which are significantly exacerbated in the conditions of a shortage of funds to finance the work.

The above is to some extent confirmed in the report prepared by the group of experts in the framework of the TACIS project "Development of Road Transport". Experts noted that "the most serious obstacle to an effective regulation of the transport industry is the absence of an integral "working" general transport policy and a sectoral policy in the field of road transport" [11].

It should be mentioned that the degree of these problems settlement determines the "maturity" of the system for ensuring road safety, "development" of the public relations in this sphere.

In different countries, the scope of "coverage" of the issues related to the road safety ranged from the consideration of individual problems to a broader analysis of the complex problems of the entire road traffic system. However, the essence of these decisions is the main thing—the recognition of the mandatory development and implementation of appropriate measures to prevent accidents and reduce the severity of RTT. Moreover, the industrialized countries have developed the legal strategy for the provision of road safety.

At the same time, the complete copying of foreign schemes, methods, approaches, and mechanisms of road safety in Russia, despite their undoubted merits, can only worsen the situation with road safety because of a mismatch of technology, methodology, and management methods with the general conditions of the transition period.

State management in the field of road safety covers a range of diverse activities.

In this case, it is interesting to explore the functions objectively conditioned by the needs of the object under control, that is, those real needs of the object that generates, necessitate the formation of the functions of the control system, i.e. contribute to the creation of management bodies.

Hence, the road traffic management can be represented as a process for ensuring the harmonization of the functioning of the components, blocks and elements of the "road traffic" system, a process in which the main task of the organization of the transportation of goods and passengers is solved along with the development and implementation of measures to optimize the functioning of the entire road traffic system.

Such optimal functioning can be achieved only if there is state control implemented on a program-target basis.

Designed in the most general form, the theoretical model of road safety makes it possible to justify the need to organize, conduct, modify (or argue in case of development) appropriate organizational and other strategic measures to solve the problem of reducing the level of RTT at every stage of the development of society and the state. The analysis shows that the country is seeking solutions of the problems of road safety in finding ways to improve management effectiveness, developing organizational management schemes with a general tendency to redistribute functions from the center to the territorial and local levels. At the same time, it can be stated that the actual dynamics of practical work to ensure the road safety is ahead of the conceptual comprehension of this problem. There is a promising, long-term task of forming a management system as a self-adjusting mechanism, based on a set of criteria defining all the necessary characteristics of road safety, conditioned by the needs of the society and the state, differing both in vertical as well as in horizontal management. The evaluation of the effectiveness of the management system can be determined through feedback, the content of which includes relevant indicators, which deviations from the accepted criteria constitute the subject of administrative influences.

The combination of the principles of centralization and decentralization for decision-making in the field of road safety should facilitate the creation of flexible, variable and simultaneously stable centralized and autonomous organizational structures. Thus, in the long term, it is about creating modular management principles that could be applied at different levels, regardless of the type and nature of the territory, and the scale of the decisions. Depending on the type of tasks to be performed, all modules, their various combinations or parts of modules can be used. The composition of the modules remains open, which reflects the condition of flexibility and customizability of the management model. The dynamic nature of the transformations in the country poses the task of the creation of flexible technological modules (which is different from shaping management technologies for specific, autonomous tasks, that is the case at present). The use of these models allows designing sets with various degree of detail that must provide a solution to various tasks: one (autonomous), a series of interrelated tasks, complex tasks to provide for the correspondence between the resource and the need.

The model proposed in the most general form for the technology of managing traffic safety can be considered as a long-term (strategic) task, requiring for its implementation a long time, detailed development, activation of a significant potential in management, financial relations, informatics, scientific research, etc.

As noted above, at the current stage of socio-economic transformation, the following tasks should be set:

– construction (in the necessary volumes) of highways and road networks in cities;
– complete renewal of the fleet of vehicles according to the design and technical conditions that meet the modern requirements in the field of road safety;
– education of the road participants with the consideration of the ethical, psychological, and traffic characteristics of the road, which might require long efforts and huge capital investment.

It is obvious, that the situation with the provision of road safety will improve with the achievement of goals aimed at political and socio-economic stability in the country. This is also confirmed by the above-mentioned authors when considering the issues of industrial safety. They note that "the permissible level of security is largely determined by the level of the development of society. And it is this level of development that limits the possibilities of applying economic, legal and organizational mechanisms". Thus, in our opinion, the effectiveness of the development and application of various road safety mechanisms directly depend on the price that society is willing to pay for its safety.

Table 1 The Haddon matrix

Phase		The system of road "traffic" (factors)		
		Individual	Automobile	Road (traffic environment)
Before a traffic accident	Prevention of traffic accident	1.1. Information, training, enforcement of laws	1.2. Admission to operation. Elements of active safety	1.3. Ensuring compliance with requirements. speed limit
Traffic accident	Prevention of injuries during a traffic accident	2.1. Use of protective equipment	2.2. Equipment providing protection against traffic accidents	2.3. Arrangement of roads, providing protection during road traffic accidents
After a traffic accident	Preservation of life	3.1. First aid skills. Access to medical care	3.2. Availability. Fire safety	3.3. Elimination and prevention of traffic "jams". Emergency care

It can also be stated that in the conditions of the unstable social and economic situation in the country, the target settings of certain programs, weakly tied to the real social and economic situation, as a rule, are not achieved.

4 Program Management in the Field of Road Safety Providing

For the development of a program to improve road safety, you can consider all the possible options and choose the best one with appropriate restrictions.

When developing models for the organization of preventive work, it seems most effective and expedient to use the approach proposed by Jr. Haddon [10].

Nine-cell Haddon matrix (Table 1) simulates a dynamic system, with each cell of the matrix providing the possibility of intervention to reduce RTT.

Various strategies to reduce the incidence of traffic accidents were developed in different countries based on the Haddon matrix. The corresponding strategies assume the activities aimed at the following objectives:

- to reduce the level of exposure to risk;
- to prevent accidents;
- to reduce the severity of injuries in the event of an accident;
- to mitigate the consequences of injury by using an improved approach after an accident.

The priority direction of activities can be determined to take into account the Haddon matrix.

Let us briefly consider possible approaches to the mechanisms for developing programs for improving road safety, regulatory and legal acts, working out the criteria for evaluating the activities of the Traffic Control Department, and ensuring the technical integrity of vehicles during operation. Improvement of activities in these areas will undoubtedly contribute to an increase in road safety, a reduction in the severity of the consequences of an accident.

As noted above, in 2003–2004s the authorities concluded that only the application of a program-targeted approach at the federal level will help to provide systematic and complex influence on the high rate of traffic accidents. Without the support at the federal level, the subjects of the Russian Federation and local self-government bodies will not be able to ensure the proper level of traffic safety in their territory.

When developing the Program, the following fundamental conceptual provisions were formulated:

1. The highest value is the life of a person. It is necessary to take all measures to ensure maximum safety of life and health of a road user.
2. Accident, death and serious trauma are no longer absolutely inevitable phenomena accompanying increased mobility of the population. These phenomena are already manageable.
3. Activities to ensure road safety—this is not the business of one agency.
4. The number of causes and factors affecting the occurrence of traffic accidents, according to experts, is several thousand. The logic should be reduced to the selection of the basic priority areas with a set of priority most effective measures.
5. It is necessary to envisage possible measures to ensure safe traffic conditions for the most vulnerable road users and, first of all, pedestrians and children.
6. It is necessary to carry out regular monitoring of traffic so that, in the event of appropriate social and economic changes in the country, make the necessary adjustments to the previously assigned tasks and implementation plans.

For the development of effective control actions in the field of the organization of road safety, it is advisable to use the mechanisms of integrated assessment. The main idea of this approach is that each vertex of the target tree is disaggregated exactly into two sublattices, that is, the so-called dichotomy method is used. This allows you to aggregate each pair of vertices to a subsequent vertex (top level) using logical convolution matrices.

In the development of the program, tasks, such as the transition of the program to a higher level of road safety, can be achieved, that is, we turn to a program with a higher integrated assessment value. Also, the task of selecting a program with a minimal risk in its implementation can be set and solved.

This methodology also allows to formulate a list of program activities, the implementation of which provides an increase in the level of security up to a given estimate.

In a formal description, the problem can be defined as follows:

Define a set of Q activities that ensure

$$\max P(Q)$$

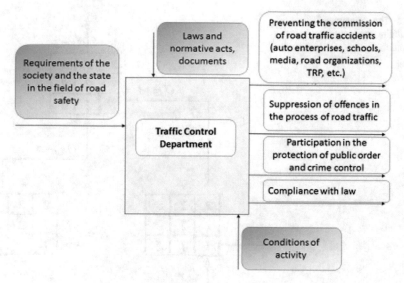

Fig. 1 The integrated model of the state automobile inspection

under restriction

$$A(Q) \geq L$$

a_i effect from the i-th event;
p_i the probability of its implementation;
Q set of program activities;
L the limit value of the cumulative number of accidents

The solution of the problem reduces to the classical problem of "knapsack", for the solution of which there are effective algorithms (the method of dynamic programming, the method of dichotomous programming, and others).

The method of integrated assessment is used to develop a methodology for assessing the performance of vehicles and divisions of the Traffic Control Department.

The evaluation of activities is carried out according to the indicators contained in the approved forms of statistical reporting. The system of indicators can be formed with the view of the solution of the problems facing the traffic police (Fig. 1).

5 Comprehensive Assessment of Road Safety Providing Effectiveness

Comprehensive assessment of Quality of Service (*QoS*) of the traffic police activities is performed on the basis of evaluations in four functional areas.

Fig. 2 The sequence of
calculating the complex
estimate

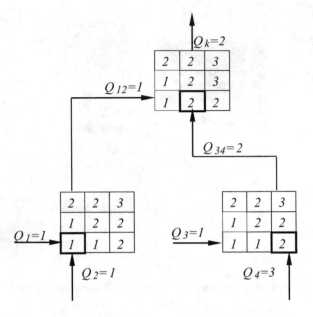

The 1st direction is monitoring compliance with traffic rules (*O1*).

The 2nd direction is the supervisory activity (*O2*).

The 3rd direction is the participation of traffic police in the fight against crime and the protection of public order (*O3*).

The 4th direction—the state of legality and official discipline (*O4*).

The pairs of directions form the so-called binary convolution structure, which illustrates the scheme of sequential obtaining, first, of generalized data, and then complex evaluation of the object.

But before forming a binary structure, a set of indicators that characterize each of the four directions is determined; while the values of one part of the indicators can be accurately calculated, to determine the values of the other part of the indicators it is required to obtain and process the opinion of experts.

After that, scales are developed for recalculating the values of indicators in the intermediate scores and scoring scale, with the help of which experts assess the figures, the exact calculation of which is impossible.

Then the importance of intermediate scores is determined expertly and pairs of directions are determined, whose local estimates will be collapsed into a generalized estimate, and a binary convolution structure is constructed. The corresponding matrices of logical convolution are formed for the constructed binary stricture, and a complex evaluation of the traffic police activities is determined.

Figure 2 shows an example of a binary convolution structure for 4 functional directions of the traffic police activities and obtaining a QR in the case of a 3-point assessment (unsatisfactory, satisfactory, good).

For the example given, we find that the complex evaluation of the traffic police activities is 2 or satisfactory.

The methodology of obtaining a comprehensive assessment using the logical convolution matrices is also applied in the evaluation of draft legal acts on road safety.

Suppose that in accordance with the methodology presented above for constructing a comprehensive assessment of the effectiveness (priority, importance, necessity, etc.) of the draft law or other normative legal act (measure), three ($n = 3$) relatively independent directions are identified:

- legal drafting of the project ($A1$);
- social demand for the project ($A2$);
- project security (procurement of resources) ($A3$).

The sequence of aggregation of assessments of directions, in this case, is determined expertly through discussions with specialists.

It is expedient to take the estimates $A2$ and $A3$ as the first pair of convoluted local estimates. This is because the assessments of these levels characterize a party oriented to the social validity of the project being evaluated. At the same time, the assessment of the level of the legal drafting of the project characterizes the degree of "maturity" of the project, the depth of analysis, legal techniques, etc.

First, the evaluation of the level of social demand and the assessment of the level of project security is aggregated into a generalized assessment of the level of social validity of the project (A_{so}). Then the assessment of the level of social validity of the project is folded with the assessment of the level of legal drafting ($A1$) in the comprehensive assessment (CA).

Figure 3 shows a scheme for aggregating estimates of these three directions into a comprehensive assessment.

When assessing draft laws in the field of road safety, the construction of logical convolution matrices makes it possible to reflect the state-power strategy, the will of the decision-making authority, because with these matrices it is easy to implement a preferred strategy in the process of assessing the projects under consideration.

If there is a whole list of draft laws, the implementation of which is aimed at solving problems in the field of road safety, then the procedure for constructing an integrated assessment is carried out for each project included in this list.

After carrying out this procedure, all the projects proposed for consideration receive their final comprehensive evaluation.

A comprehensive evaluation of each draft law can be considered a priority of this law. With this approach, the higher score received by the project corresponds to its higher priority. If there is still an assessment of the funds that should be spent on the implementation of each project, then it becomes possible to streamline the existing list of projects and determine the sequence of implementation of laws based on their comprehensive assessments, that is, to form a hierarchy of analyzed projects.

The efficiency of the project is defined as a quotient from the division of the integrated assessment into costs, that is,

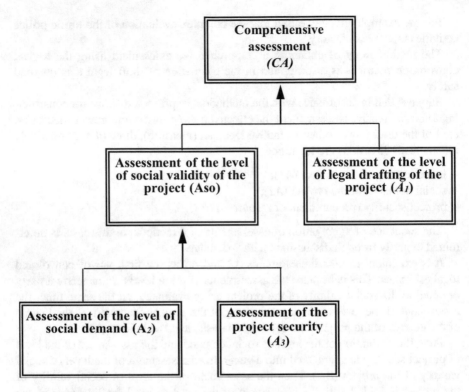

Fig. 3 Scheme of aggregating estimates in three directions

$$E_i = \frac{I A_i}{C_i}, \quad i = 1, \ldots, 5.$$

In fact, it defines the efficiency of using funds, i.e. what effect the society receives from one ruble spent on implementing the law.

It is easy to visualize the effectiveness of the project in the form of a graph if you set the cost values on the abscissa axis and the values of complex estimates along the ordinate. In this case, we obtain a bundle of segments emerging from the origin (Fig. 4).

6 Effectiveness Maximization Problem

To get the maximum effect, the most efficient project is executed first, then the next one inefficiency, etc. Figure 4 clearly shows in what sequence it is necessary to implement projects.

From this figure it follows that the fourth project is the most effective, then the third project, then the second, the first and, finally, the fifth project.

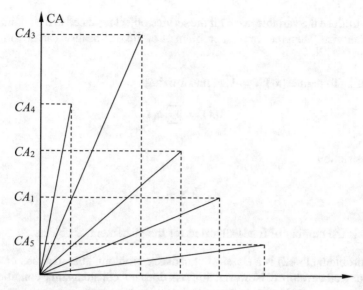

Fig. 4 Effectiveness of projects

The resulting values of the efficiency allow us to construct a "cost-effect" schedule, from which, in addition to the sequence of projects, it is clear what maximum effect can be obtained from the implementation of these projects, and what resources must be invested in the implementation of these projects.

Thus, if the financial component of the draft laws is taken into account, the sequence of their implementation will be significantly affected by the costs associated with their implementation.

In solving problems related to the provision of vehicles in a technically sound condition, there is some interest in both the organizational model for carrying out technical inspection of vehicles and the task of optimally locating the network of stations and technical inspection points.

The task of placing objects of various types comprises a wide class of problems of discrete optimization. Let us sort out the tasks of stationing the network of stations and points of the technical inspection. There are various options for setting the tasks for the optimal allocation of stations and points of the technical inspection depending on which constraints are significant and which optimality criteria are chosen. Let us consider some of these options.

Let n be points for the possible location of the network of technical inspection stations—further objects. We assume that all objects are of the same type in the sense that the effect of their placement depends only on the placement point.

We denote by a_i—the effect of the functioning of the object in point I; b_i—the cost of its placement and commissioning in point i.

We introduce the variables $x_i = 1$ if the service object is placed in the i point, and $x_i = 0$—otherwise. Then the simplest problem of optimal allocation can be formulated as follows.

Problem 1 To define $\{x_i\}$, $i = \overline{1,n}$, maximizing

$$A(x) = \sum_i a_i x_i \tag{1}$$

under restriction

$$\sum_i b_i x_i \leq B, \tag{2}$$

where B is the number of funds allocated for the placement of objects.

The problem (1)–(2) is a classical "knapsack problem", the methods of solving which are well developed. However, this task does not consider some conditions that may prove to be significant. In fact, placing a lot of objects in one region reduces the effect of each of them due to the limited needs of the population in this type of service.

For example, if all items of the possible location of objects are situated in one region, then the corresponding restriction has the form

$$\sum_i x_i \leq p,$$

where p is the maximum number of objects that it is advisable to place in a given region.

If there are several regions, and in the k-th region there are a lot of p_k possible locations for objects, we get the following system of restrictions

$$\sum_{i \in p} x_i \leq p_k, \quad k = \overline{1,r}, \tag{3}$$

where p_k is the maximum number of objects that it is advisable to place in the k-th region, and r is the number of regions.

In some cases, the condition of non-placement of two objects in close or neighboring points is essential. The closeness of the points is conveniently given in the form of a graph whose vertices correspond to the placement points, and the edges connect the neighboring points.

If U is the set of edges of the graph of neighborhood points, then the restrictions associated with the non-placement of two objects in neighboring points take the form

$$x_i + x_j \leq 1, \quad (i, j) \in U \tag{4}$$

Note that if the restriction on the number of financial resources is not significant, the problem (1), (4) is the task of determining the independent set of vertices of the graph having the maximum sum of weights a_i.

Comment. When setting the object placement task, it is assumed that the placement of the same type of objects belonging to other structures is known, which allows us to estimate the expected effect from the placement of objects.

Problem 2 To determine $\{x_i\}$, $i = \overline{1, n}$, maximizing (1) under the restrictions (2) and (3).

Consider generalizations of problems 1 and 2. Let objects not be of the same type. In this case, both the effect and the cost of placing the object depend both on the type of the object and on the placement point.

Denote, respectively, a_{ij}—effect, b_{ij}—costs, if the object of the i-th type is located in the point j. We introduce the variables $x_{ij} = 1$ if the object of type i is located at the point j, and $x_{ij} = 0$—otherwise. Let the number of object types be m.

Problem 3 To denote $\{x_{ij}\}$, $i = \overline{1, m}$, $j = \overline{1, n}$, maximizing

$$A(x) = \sum_{i,j} a_{ij} x_{ij}$$

under restrictions

$$\sum_{i,j} b_{ij} x_{ij} \le B$$

$$\sum_i x_{ij} \le D_j, \quad j = \overline{1, n}, \tag{5}$$

Restrictions (5) reflect the fact that at each point you can place no more than D_j objects of different types. So, for example, at one point it is possible to place a point of the technical inspection and a cafe (and possibly a hotel), which may prove to be the most effective.

Problem 3 does not take into account the fact that when placing objects of different types in one point, the effect is usually greater than the sum of the effects when placing these objects without considering their joint functioning, and costs are usually less than the sum of costs for an independent location (there is a so-called synergistic effect).

With the view of these peculiarities, we'll proceed as follows. We will consider a complex consisting of one or several objects of different types as an object of a certain type.

So, for example, we'll take a complex, consisting of the stations of technical inspection, car service, and cafes. This approach allows to consider the synergistic effect, although the number of types of objects increases. In this case, in the restriction (5) of Problem 3, we must put all $D_j = 1$, since we can place no more than one complex at one point.

7 Conclusion

Consideration of the restriction of the form (3) in problem 3 is more complicated since we are talking about the functioning of complexes of different types.

Assume that in each complex there is a defining type of object, and all other objects entering the complex are complementary.

For example, if the station (point) of the technical inspection is the determining object, then the car service and the cafes included in the complex are complementary. They are aimed at customer service, which gives a synergistic effect. This approach permits to take into account the constraints of the form (3) only by the defining type of objects, which greatly simplifies both the formulation and the solution of the problem.

Indeed, in this case, all complex objects (complexes) are broken into non-overlapping classes according to the defining type of objects, and constraints of the form (3) are written out for each class of objects.

The provisions proposed above have largely been reflected in practical activities to improve road safety including the development and implementation of federal targeted programs.

As it is known, during the implementation period of the program of 2006–2012s, the total number of deaths in road accidents decreased by 19.0%, pedestrians' deaths by 43.0%, children's deaths by 34.0%. For 3 years of the implementation of the program of 2013–2020s, the number of deaths has also decreased.

References

1. Road traffic injuries: http://www.who.int/violence_injury_prevention/road_traffic/en/. Access date 1 July 2018
2. State report "On the State of Road Safety in the Russian Federation", 142 p. (2007)
3. Hauer, E.: The Art of Regression Modeling in Road Safety, 233 p. Springer International Publishing (2013)
4. Manneringa, F.L., Bhatb, C.R.: Analytic methods in accident research: Methodological frontier and future directions. Anal. Methods Accid. Res. **1**, 1–22 (2014)
5. Golts, G.A:. The trend of change and the forecast of the role of motor transport in urban passenger transportation in Russia. Socio-economic problems of development of transport systems of cities and zones of their influence. In: Materials of the VIII International Scientific and Practical Conference, USEU, pp. 19–23 (2002)
6. Bhat, C.R., Born, K., Sidharthan, R., Bhat, P.C.: A count data model with endogenous covariates: Formulation and application to roadway crash frequency at intersections. Anal. Methods Accid. Res. **1**, 53–71 (2014)
7. Bondarenko, Yu.V., Azarnova, T.V., Kashirina, I.L., Goroshko, I.V.: Mathematical models and methods of assisting state subsidy distribution at the regional level. In: International Conference "Applied Mathematics, Computational Science and Mechanics: Current Problems", Voronezh, Russian Federation, 18–20 Dec 2017. https://doi.org/10.1088/1742-6596/973/1/012061
8. Ye, F., Lord, D.: Comparing three commonly used crash severity models on sample size requirements: multinomial logit, ordered probit and mixed logit models. Anal. Methods Accid. Res. **1**, 72–85 (2014)

9. Amalberti, R.: The paradoxes of almost totally safe transportation systems. Saf. Sci. **37**(2–3), 109–126 (2001)
10. Haddon Jr., W.: The changing approach to the epidemiology, prevention, and amelioration of Trauma: the transition to approaches etiologically rather than descriptively. Am. J. Public Health **58**, 1431–1438 (1968)
11. Legal Regulation of Motor Transport Activities: Report. The TACIS project "Development of Road Transport", p. 9 (1999)

Big Data-Driven Control Technology for the Heterarchic System (Building Cluster Case-Study)

Dmitry Anufriev, Irina Petrova, Alla Kravets
and Sergey Vasiliev

Abstract This chapter presents a new approach to describing control mechanisms in the regional building cluster which is viewed as a heterarchical system. The structural model of building cluster, based on this description, is represented. The heterogeneity of the information environment caused by the fact that the entities as are united in a cluster of the primary production branch, and the accompanying industries, complicates the creation of the common information space supporting activities of a cluster. The functional and information simulation of the system is executed for the purpose of formation of the integrated information and communication environment on the example of clusters of several regions of Russia. As a result of the fulfilled research regression models were built for the short-range forecast of values within the framework of analysis of building cluster. The forecast of volumes of input of accommodation in region was chosen as a task. The built regression models will be able adequately to do prognoses on a short period.

Keywords Cluster · Heterarchical system · Holon
Multivariable regression models · Short period prediction

D. Anufriev · I. Petrova (✉)
Astrakhan State University of Architecture and Civil Engineering, Astrakhan 414056, Russia
e-mail: irapet1949@gmail.com

D. Anufriev
e-mail: adp_2000@mail.ru

A. Kravets · S. Vasiliev
Volgograd State Technical University, Volgograd 400005, Russia
e-mail: agk@gde.ru

S. Vasiliev
e-mail: svasilev2012@yandex.ru

© Springer Nature Switzerland AG 2019
A. G. Kravets (ed.), *Big Data-driven World: Legislation Issues and Control Technologies*, Studies in Systems, Decision and Control 181,
https://doi.org/10.1007/978-3-030-01358-5_18

1 Introduction

Cluster approach allows increasing the effectiveness of the work and development of regional building complexes, and, consequently, and the economy as a whole [1]. A regional building cluster (RBC) allows to increase the quality of building products, bring down costs, improve housing affordability for people, and also to create conditions for the expansion of scientific research-and-developments (R&D) in the building. However, in spite of plenty of foreign and Russian researches, management problem of business processes in an RBC has not been completely solved yet.

A heterarchy is a new method of organization, being neither market nor hierarchical. While hierarchies presupposes strict subordination over levels and markets suppose relations of complete independence, heterarchies presuppose the relations of interdependence. As follows from the term, heterarchies are characterized by the minimum degree of hierarchy and organized heterogeneity [2].

An association of participants in a cluster is the socio-economic system (SES) functioning on the basis of the institutional mechanism of coordination. Its forming supposes the presence of connections between participants by virtue of geographical closeness, and also the presence of partial cooperation on the basis of market mechanisms. Moreover the weakly structured connections of economic cooperation in SES, increasing complexity and novelty of management tasks in conditions of insufficient information, stochastic environmental impact increase the uncertainty factor in the acceptance of administrative decisions. Thus, a cluster is the new form of organization of SES, based on heterarchical approach. Cluster management strategies require the creation of the integrated information support mechanisms for decision-making processes.

Nowadays the cluster approach is widely used near segmenting of regional economy for the achievement of global aim—to receipt of maximum profit conducive to the development of each participant of the cluster and the region as a whole. The formation and development of national and regional clusters responds both to the macro-needs of the national economy, and to the micro-needs of entrepreneurial structures.

2 Background

A presence in the structure of a regional cluster of the developed infrastructure of service, consultative services, suppliers, will allow to bring down costs and to increase the competitiveness of any large firm. In this case, the role of the cluster's core is played by large companies, but small and midsize businesses developing around them at an accelerated pace become an important payer in a budget and basic source of development of the territory. As a single integration and diversified mechanism, a cluster will not only bring down costs of production but also will liquidate uncompetitive duplication.

Effective clusters are an object of intent attention and large capital investments of governments both in Europe and in Russian Federation. Concentrating efforts on supporting existing clusters and creating new networks of companies that had not previously contacted each other, regional governments not only promote the formation of clusters but also become participants in cluster networks themselves.

In 2016 the Ministry of economic development of the Russian Federation started a priority project "Development of innovative clusters-leaders of investment attractiveness of world level". This project embraces the wide range of hi-tech industries: from pharmaceutics to aerospace industry, and had been implemented in 11 regions of Russian Federation. Today it is possible to distinguish the next groups of clusters: innovative, industrial, tourist, agro-industrial. These clusters participate both in federal and regional programs of support under the budgetary financing. Thus according to data of the Russian cluster observatory State Research University HSE only one cluster from 115th belongs to specialization "Building, municipal economy, architecture and technical tests", and only one program among 11 programs of innovative clusters-leaders(consortium of innovative clusters of Moscow region) is indicated in the direction "Building technologies" [3].

At the same time in Europe, construction clusters are well represented. In the analytical report of European Commission for 2017 [4], it is marked that a building sector plays a strategic role in the economy of Europe (EU) because it gives 9% gross domestic product (GDP) and 18 million of workplaces. Building clusters were presented in 19 countries (Portugal, Slovakia, Finland, Italy, Norway, Spain, Greece, Lithuania and other).

Thus, imperfection of existent formal methods and instruments of organization and development of the heterarchical system of regional cluster gives results not only in technological lag of separate institutional industries (Fig. 1) but also informing of scientific and technical challenge: imperfections of theoretical bases of automated management and support of acceptance of administrative decisions, that implement the strategy for the development of the heterarchic system of the regional cluster at all stages of its life cycle, in a single information space of knowledge about objects, processes, environments, decision-making procedures.

3 RBC as a Heterarchical Socio-Economic System

In the economic complex system, a leading role is played by the process of production (without which the processes of distribution, exchange, and consumption are impossible) of material welfares, characterized by certain economic indicators.

Passing to the market relations in housing policy fundamentally changed the role of the state in building industry and housing services. In the housing market, private construction companies, companies providing housing and communal services, solvent housing buyers and consumers of housing and communal services interact with each other. In the conditions of market economy, the possibilities of regulation

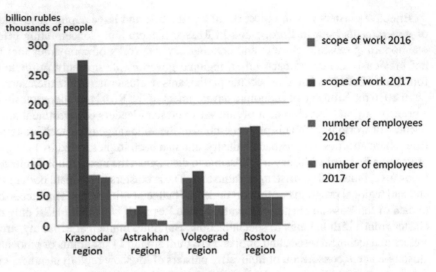

Fig. 1 Socio-economic indicators of building industry

of building industry have a limited character in the part of government bodies and municipalities.

Using the concepts of new economy of the adaptive systems [5], it is possible to say, that the problem of any transformed economy consists of capacity for adaptation in the change of organizational structure so that she strengthened the ability to react on unforeseeable future changes in an environment [6, 7].

For the development of the economy of the region, it is necessary to apply of such form of organization and co-operation of labor that would provide an accumulation and effective use of resources of the territory. It is possible to use clusters for such forms.

Founder of cluster approach, M. Porter, offered next determination of cluster: it is a group of geographically neighboring associate companies (providers, producers) and organizations (educational establishments, government bodies, infrastructural companies) related to them, operating in a certain sphere and complement each other [8].

Thus, it is possible to offer the next determination: the RBC is regionally localized interconnected complementary enterprises of construction and related industries, united by heterarchic relations with local institutions, authorities, cooperating enterprises with the aim of increasing the competitiveness of these enterprises and the regional economy as a whole.

M. Porter [8] considered that clusters could have an influence on a competition by three methods:

– increasing of the productivity of enterprises—participants of the cluster;
– stimulation of innovations;
– stimulation of development of new business-directions.

The cluster analysis of the socio-economic system pursues next aims:

– determination of basic directions of development of cluster: basic and supporting industries;
– structuring of resources;
– determination of the underlying structure of cluster;
– determination of environment of direct and indirect cooperation;
– determination of the strategy of development of cluster.

For the determination of the structure of RBC the next types of researches were used:

1. research of logistics of material, financial and competences' streams;
2. analysis of information flows in the system of cooperation of imperious, productive, scientifically-educational and innovative structures in the building complex of the region;
3. analysis of processes of decision making, to their apartness or connectivity on the basis of principles of teamwork in decision-making;
4. analysis of the distribution of power, responsibility, teamwork in decisions making.

Based on researches [1] four groups of basic participants of building cluster were distinguished:

1. Governmental structures at the level of the region, municipalities.
2. Large companies in the industry.
3. Companies interrelated with regional branch vendors.
4. Public structures.

Creation of building cluster will allow to optimize the contacts of branch enterprises, put right effective co-operation, co-ordinate the plans of enterprises carrying out the different types of economic activity.

At the same time, the efficient functioning of cluster requires the creation of unified information space, within which a dialogue will be conducted between business, government, and education about the ways of development of competitive edges within the framework of the region and country. As a result, cluster cooperation will allow operatively adapt underlying structures and external intercommunications to the quickly changing environment.

Scientific novelty of base conception of genesis and evolution of mechanisms of the automated management in heterarchical SES consists of that management principles which are first formulated and reasoned in this class of the systems on the basis of analysis of data about the main types of administrative, financial, material and skilled resources on all stages of life cycle of RBC, that allows to realize the complex of economic, informative and cognitive managers of influences.

Presently a building complex in the vast majority of regions of Russian Federation is uncertain essence [1] that consists of a great number of independently managing objects, not having the centralized cooperation between building enterprises, administration of area or country, educational establishments, and having individual features.

A building cluster is the concentrated group of the organizations related to the building (higher educational establishments, public organizations, companies on the production of building materials, company on providing of building services), and their interaction give afterward certain advantages of other separate building companies.

In a building cluster, a social effect from a clustering is expressed in the increase of the level of availability of accommodation for a population. In those regions where building clusters are formed, house construction volumes are much larger. Presence in the cluster of attendant elements as banks, legal government control, the stability of the system of the cluster allows promoting this index through providing of population accessible mortgage loans with the assured terms of input of accommodation in exploitation.

Thus, it is necessary to examine clusters as a new method of the aggregated use of advantages of branch location of organizations and possibilities of regional management. Clusters are oriented to the economic success of territory of their location and, like the authorities that are entrusted with the management of the region or the state, can provide economic development, bringing in of additional investments, activation of innovative processes, and also the solution of many social problems.

4   Structural Model of RBC

The main feature of clusters is a network form of organization of the productive cycle. An economic effect from the creation of clusters in a building sphere is conditioned [1]:

(1) by a productively-building cooperation allowing effectively use the combined potential of network partners;

(2) by a cost cutout on the modernization of building products by the transmission of part of works to the partners specialized in the certain types of activity;

(3) by the increase of efficiency of the process of providing of building production raw and other materials, details, constructions on the basis of establishing long-term partner connections;

(4) by the increase of efficiency of implementation of separate administrative functions due to the division of labor, specialization, bringing in of the specialized organizations of building a profile;

(5) by the increase of efficiency of works in the area of sale and service, acquisition of necessary resources;

(6) by the increase of reliability of network partners in investment-financial cooperation.

Normal development and functioning of building complex depend, foremost, from balanced of investment demand, material and technical development of this sphere, presence, and efficiency of work of institutes of competition, and, in the final analysis, from the efficiency of government and regional control and self-regulation by the activity of subjects of the building complex.

But it requires the use of new approaches and methods of the decision of nascent problems and tasks of changes in structures and models of the behavior of building organizations.

Cluster conception of gilding the development of enterprise in the field of building focuses on intercommunications between corporate structures, investment, intermediary, scientific, educational, public organizations of the region.

Advantages of association in a cluster [1]:

1. The possibility of bringing in of financial resources in a building sphere—(by the association of general financial possibilities of participants of the cluster, bringing in of investments, joint participating in the competitions of projects that is financed as grants, an association of general financial possibilities for providing of guarantees on the receipt of credit resources).

So, an association of financial resources of participants of the cluster will give an opportunity on the base of one of the participants of the cluster to organize a productive line on producing building materials. It will give an opportunity to all participants of the cluster to pass to the use of own building materials.

2. The possibility of the decline of building unit and services of the organizations included in a cluster cost.

Presence of own production on the basis of local source of raw materials assists a cost cutout on transporting, maintenance, trade services, that reduces in price the prime price of dwelling-place and repair works in turn.

3. To provide the permanent work-load of the organization of building industry and contiguous industries; creation of new workplaces; assistance to the revival of the industry of separate districts of the area.
4. An association of organizations at the level of the region is a fundamentally new level of relations, that is based on decency and trust between the participants of the cluster and it is possible to say the new way of thinking.

For example, today the union of building companies, that links more than 300 firms of city and area, is realized in the city of Saint Petersburg [9]. This union exists more than six years and continues consistently to increase, that shows the efficiency of his use. The fact of creation of building cluster shows his efficiency, thus, a building cluster was also realized in the next areas: the Voronezh area, Kaluga area, Novosibirsk area and Sverdlovsk area.

The structural model of building cluster, based on description higher, is represented on a Fig. 2.

This model shows the cooperation of separate structures (enterprises, institutes, government bodies), using necessary instruments for the achievement of eventual product development.

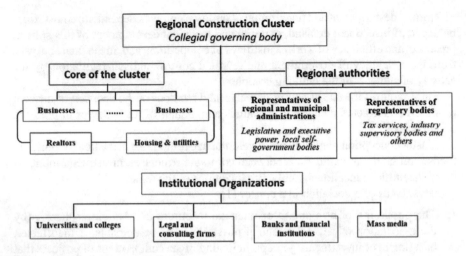

Fig. 2 Model of the building cluster structure

5 Forecasting of RBC Development

5.1 Determination of Attributes Set for Real Estate Market and Data Processing

A real estate market is the system of interrelated market mechanisms providing creation, transmission, exploitation, and financing of objects of the real estate [10].

Based on the definition, a subject domain includes the great number of parameters, intercommunication between which can be identified by calculating the coefficient of correlation or on the basis of expert opinion.

In this chapter, emphasis is placed on the construction of regression models for prognostication, i.e. without the detailed selection of certain attributes (the selection of which depends on a task).

At the same time, the data under consideration, for the most part, cover the regional level, since a real estate market is localized, and every region has the specific conditions.

As an example let us choose the most common task of prognostication of housing commissioning volumes in a region (1000 m^2 of general area) as the target (dependent) indicator.

The set of attributes is chosen empirically (on the basis of work [10])

(1) The population size (end-of-year estimation), thousand persons;
(2) Average per capita monetary income of population, rubles;
(3) Number of operating building organizations;
(4) Average prices at the primary housing market, rubles per square meter of the total area;

(5) Average prices at the secondary housing market of rubles per square meter of the total area;
(6) Gross regional product: per capita, rubles;
(7) Investments in the construction industry, millions of rubles;
(8) The volume of the works, executed by appearance to activity "Construction", millions of rubles;
(9) Dollar-ruble exchange rate.

Set of statistics is compiled mostly from data of Rosstat [11] and presented in the context of following federal entities of Russian Federation:

– Moscow;
– Astrakhan region;
– Volgograd region.

Experimental data are processed and transformed into a file with a .csv format that contains the following data:

– implementation—a commissioning of dwelling-houses;
– people—a population;
– builders—construction organizations;
– finance—average per capita monetary income of population;
– dollar-ruble exchange rate;
– primary_housing—an average cost of sq.m of the primary housing;
– secondary_housing—an average cost of sq.m of the secondary housing market;
– workload—a volume of works to activity "Construction";
– grp—a gross regional product;
– investment—investments in the construction industry.

When loading data on the one region for training is selecting separately.

5.2 Evaluation Methods of Regression Models

The aim of experiments is a developing of regression models for prognostication of data in the domain of construction industry and real estate market. The emphasis is placed on a regional level leading to unique conditions of markets and entailing the corresponding selection of data.

For a selection the following types of regression models are taken:

(1) Linear regression model;
(2) Regression model on the method of support vectors (SVM);
(3) A regression model of decision trees.

As a test dataset, we will take the last two values of the temporal series. On the method of the fixed zero point ("Fixed origin") we will clean the last $h = 2$ inquiries from a selection (for 2014 and 2015 years) and build a model by the first T to the h values, whereupon check exactness of models by means of following coefficients:

(1) Mean Absolute Error (MAE):

$$\frac{1}{n}\sum_{i=1}^{R}|Actual - Predicted| \tag{1}$$

(2) Mean Squared Error (MSE):

$$\frac{1}{n}\sum_{i=1}^{R}|Actual - Predicted|^2 \tag{2}$$

(3) Root Mean Squared Error (RMSE):

$$\sqrt{\frac{1}{n}\sum_{i=1}^{R}|Actual - Predicted|^2} \tag{3}$$

(4) R-squared (R^2):

$$R^2 = 1 - \frac{SS_{res}/n}{SS_{tot}/n} = 1 - \frac{SS_{res}}{SS_{tot}}, \tag{4}$$

where $SS_{res} = \sum_{i=1}^{n} e_i^2 = \sum_{i=1}^{n}(y_i - \hat{y}_i)$ is a residual sum of squares of regression;

$SS_{tot} = \sum_{i=1}^{n}(y_i - \bar{y})^2 = n\hat{\sigma}_y^2$ is a lump sum of squares.

The coefficient of determination is considered as a universal measure of the dependence of one stochastic value on the set of others. The closer it gets to 1, the closer built equalization to the selection. A calculation is based on the training part of the selection.

For the additional evident estimation of prediction of resulting models, we will take an attribute vector according to federal entities for the 2016 year (see Table 1). Missing statistical data are replaced by the arithmetic mean for the last 3 years for the entity of the Russian Federation.

5.3 Regression Models for the Assessment of Housing Commissioning

Many models become inapplicable in the conditions of the uncertainty of socio-economic processes, increasing the instability of mechanisms of economic development (especially taking in account crises or political-economy limitations), change-

Table 1 Test vectors attributes

Attributes	RF regions		
	Moscow	Astrakhan region	Volgograd region
People	12748.17	1018.24	2508.26
Builders	101490.77	1844.61	4725.95
Finance	63558.72	23999.56	23198.68
Dollar	59.5965	59.5965	59.5965
Primary_housing	164315.44	41601.61	40553.64
Secondary_housing	172898.64	41289.56	39702.36
Workload	960126	46048.81	89209.4
grp	1243600	327684.36	296255.42

ability of industrial and economic relations, variabilities of the majority of the social and economic process [12]. Therefore, there is one of the basic problems of forecasting the indexes of investments in the housing construction, as by their nature they appear most vulnerable to inflationary fluctuations, demonstrate high mobility and elasticity to the changes in demand of housing and construction industry services.

In modern conditions of socio-economic systems operating, the problem of receiving the acceptable prognosis can be solved due to combining classic methods together with the methods of intellectual prognostication, realized on principles of computer-aided instruction.

In the chapter, authors research possibilities of a few models of prognostication: linear regression, a method of support vectors (Supported Vector Machine—SVM) and method of decision trees.

Model of linear regression

The model of multivariable linear regression is:

$$Y = \beta_0 + \beta_1 \cdot 1 + \beta_2 \cdot 2 + \beta_n \cdot X_n \tag{5}$$

The parameters of every X_n model are influenced by a least-squares method in training selection.

We will train a linear model on data by regions. In addition, we will check what attributes are most meaningful using the method of Recursive Feature Elimination (RFE), consistently deleting attributes from a selection and calculating exactness of model.

Designing conclusion and interpretation by regions below.

The coefficients of regression for the models of different regions are presented in Table 2.

Using the coefficients of regression, it is similarly possible to estimate the meaningfulness of attributes. Most weight modulo has attributes of the quantity of population, the dollar rate and a number of builders. However, the data contain differences for regions.

Table 2 Coefficients of linear regressions

Attributes	RF regions		
	Moscow	Astrakhan region	Volgograd region
People	−1.380984	282.901019	−23.661646
Builder	2.369845	14.739092	−0.183064
Finance	−1.876742	−0.549627	−0.180442
Dollar	547.297794	6.694827	2.085317
Primary_housing	0.393690	−0.322464	0.092935
Secondary_housing	−0.696481	0.064210	−0.051584
Workload	−0.085680	0.022768	−0.035044
grp	0.005930	0.011565	0.012038

Table 3 Metrics of regression linear models

Metrics	RF regions		
	Moscow	Astrakhan region	Volgograd region
MAE	20081.63	1392.36	841.40
MSE	434697633.51	1951267.87	708081.88
RMSE	20849.40	1396.87	841.47
R2-Score	1.0	1.0	1.0

Table 4 Test vector for a linear model

Entering housing	RF regions		
	Moscow	Astrakhan region	Volgograd region
Forecast	−824.52	4765223.83	1346.06
Expected (trend)	2484.52	488.12	531.86
Absolute error	1660	4764735.71	814.2

The metrics of relative estimation of regression models are presented by a Table 3.

The root-mean-squared error of MAE considerably exceeds 10% of the mean value of housing commissioning for the forecasting period, and the coefficient of determination is inadequately equal to 1.

The results of predictions of models on a test vector are presented in Table 4.

The charts of rests (error of divergence) for Moscow, Volgograd, and Astrakhan regions are presented in Fig. 3.

Based on the obtained data, it is possible to say, that the models of multivariable linear regression in that kind as they presented inadequately describe the real data. On the other hand, maybe, the set of attributes is badly chosen for this task.

Model of regression by the method of support vectors (Support Vector Machine—SVM)

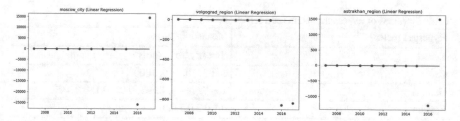

Fig. 3 Charts of bits and pieces of the linear model

The machines of support vectors (SVM—Support Vector Machine) fall into the category of universal networks of direct distribution, as multi-layered perceptron and networks on the basis of radial base functions.

The advantage over other methods of regression analysis is a determination of parameters of the regression model by the decision of task of the quadratic programming, having one decision only. The machine of support vectors can provide a high quality of generalization, not possessing a priori knowledge about the subject domain of a certain task.

When SVM is used, nonlinear regression in initial space of F can be examined as a linear regression construction problem in some extended space of H, that has a greater number of dimension than F.

The initial information for performing regression analysis on SVM is a training selection of the form $S = ((x1, y1), (x2, y2),…,(xn, yn))$. Then the task of linear regression is finding the function:

$$f(x) = (w*x) + b \qquad (6)$$

where **w** is normal to the hyperplane vector.

Functional dependence is searched by training of the SVM model on a dataset.

The construction of the algorithm for SVM training is based on the concept of the kernel of the scalar product of the support vector and the vector taken from the input space. In the model for the basis of calculations the following kernels can be chosen: linear, polynomial (poly), radial base functions (RBF) and sigmoid. In practice, the kernels are selected for which a corresponding surface separates a training selection in the best way. Today RBF is the most popular type of kernel, used in the method of support vectors.

The adjustable parameters of the model are:

(1) The margin of error C (controls a compromise between the smooth border of a decision making and correctly distributes training points);
(2) Parameter γ (gamma) is a coefficient of the kernel for 'RBF', 'poly' and 'sigmoid' kernels (the higher gamut value, the more precisely model will try to correspond to the set of data);
(3) Perimeter ε (epsilon) is a coefficient of error rounding off.

Table 5 Values of hyperparameters for a selection

Hyperparameter	Means
Kernel	'linear', 'rbf', 'poly'
C	1, 1.5, 10, 100, 1000
Gamma	1e−7, 1e−4, 1e−3, 0.01, 0.1, 0.2, 0.5
Epsilon	0.1,0.2,0.5,0.3

Table 6 Results of selection of a model of SVM on regions

RF region	Model parameters
Moscow	Default model: C = 1e3, epsilon = 0.1, gamma = 0.1, kernel = rbf
Astrakhan region	GridSearchCV: C = 1000, epsilon = 0.5, gamma = 1e-07, kernel = rbf
Volgograd region	GridSearchCV: C = 100, epsilon = 0.5, gamma = 1e-07, kernel = rbf

Table 7 SVM models metrics

Metrics	RF region		
	Moscow	Astrakhan region	Volgograd region
MAE	988.15	127.0	220.52
MSE	1430873.70	20702.20	78863.40
RMSE	1196.19	143.88	280.82
R2-Score	0.81	0.99	0.0427

Table 8 SVM models prediction on a test vector

Housing constructions	RF region		
	Moscow	Astrakhan region	Volgograd region
Forecast	3070.96	543.61	799.89
Expect (trend)	2484.52	488.12	531.86
Absolute error	586.44	55.49	268.03

For the effective selection of hyperparameters of the model the grid search method is used, consisting of the exhaustive search of possible values and estimation of the best combination. A computational experiment with the use of a model of SVM was realized in Python programming language. Scikit-learn Library has the built-in realization of this method (GridSearchCV). Variations of hyperparameters are presented by a Table 5.

The depth of retro-forecast is 2 years. Initial attributes were not excluded.

The result of the selection the best model parameters by regions is presented in Table 6.

The metrics of estimation the SVR models are presented in Table 7.

The results of predictions of SVM models on a test vector are presented in Table 8.

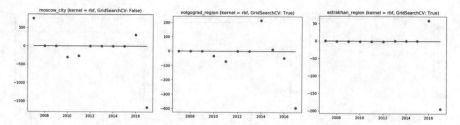

Fig. 4 Charts of rests of a model of SVM

Table 9 Metrics of the DT model

Metrics	RF region		
	Moscow	Astrakhan region	Volgograd region
MAE	988.15	127.0	332.25
MSE	1874291.425	25930.0	140753.125
RMSE	1369.047	161.02	375.17
R2-Score	0.84	0.99	1.0

The charts of rests in Moscow, Volgograd, and Astrakhan regions are presented by Fig. 4 accordingly.

By reference to the obtained results, it is possible to say, that the model much better describes data on regions as compared to a linear model. For all sets of data, the best kernel of an SVM model is RBF. Prognosis values for the examined federate entities some differ from the real ones because of sharp enough changing of dynamics for 2017 in all regions. This problem can be related to the selection of meaning attributes. Prognosis values in 2018 are some overpriced but taking into account the specifics of the formed set give an adequate enough result.

Decision Trees

A tree of decision making (Decision Tree—DT) is a tree, the values of objective function stand in the leaves of one, and in the other knots—transition terms identifying which of the ribs to go. In regression, some value of the objective function is specified in knots.

On test results the optimal maximum tree length (k) is chosen for data sets:

- Moscow: k = 2;
- Volgograd region: k = 5;
- Astrakhan region: k = 5.

Metrics of the DT model are presented in Table 9.

Judging by the coefficient of determination the decision trees describe a model on test data well enough, however it is not the indicator of forecasting quality.

The charts of rests in Moscow, Volgograd, and Astrakhan regions are presented in Fig. 5.

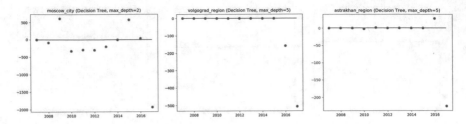

Fig. 5 The charts of rests of decision tree models

Judging on predicted values for a period it is possible to note, that for the regression model of decision trees a divergence is similarly noticeable in the strong dynamics of objective value, as well as in previous models.

5.4 Models Evaluation

Based on the comparison of MAE error value (calculated on a test set, i.e. two values for 2016 and 2017 prognosis) and module of difference of predicted and real vector (for 2018) it is possible to choose the most successful model for prediction of dataset - SVM (given in bold in header of Table 10).

It is possible to say, that all datasets are much more successfully described by a regression model on the base of the method of SVM.

6 Conclusions

As a result of the research, regression models were built for prediction of values within the framework of analysis of the construction industry cluster. As a task, the prognosis of housing commissioning volumes by regions was chosen (purpose index).

As multivariable regression models a linear model, models based on the method of support vectors of SVR and trees of a decision making were used.

Training data included statistical datasets on three regions of the Russian Federation (Moscow, Volgograd, and Astrakhan regions) for the period 2007–2017. Test data selected from the 2016–2017 period, i.e. the depth of retro prediction was 2 years.

The estimation of models was produced by value middle absolute error MAE, the coefficient of determination (R^2) and test vector of values for every set of data for 2018.

When comparing models, it is determined that for all data sets the SVR model is better suited.

Table 10 Evaluation results

Region/Metrics/Model	MAE	MSE	RMSE	R2-Score
Linear				
Moscow	20081.63	434697633.51	20849.40	1.0
Astrakhan region	1392.36	1951267.87	1396.87	1.0
Volgograd region	841.40	708081.88	841.47	1.0
SVM				
Moscow	988.15 (33.32%)	1430873.7	1196.19	0.81
Astrakhan region	127.0 (23.90%)	20702.20	143.88	0.99
Volgograd region	220.52 (28.59%)	78863.40	280.82	0.0427
DT				
Moscow	988.15	1874291.425	1369.047	0.84
Astrakhan region	127.0	25930.0	161.02	0.99
Volgograd region	332.25	140753.125	375.17	1.0

Test vector 2018		Moscow		Astrakhan region		Volgograd region	
	Forecast	−824.52	3070.96	3344.4	543.61	1346.06	799.89	912
	Error	1660	586.44	859.88	55.49	814.2	268.03	380.14
	Trend	2484.52	488.12		531.86			

In current status the models give a rudely enough prognosis: MAE for the datasets of Moscow, Astrakhan and Volgograd in the percentage of the housing commissioning value for 2007–2017 are 33.32, 23.90 and 28.59 accordingly. It is thus necessary to take into account that a small part of missing statistical information was replaced by averages values, and a test vector (values of independent variables of regression) was made on trends. However, on the whole, the tendency of data definition is retained by modeling and at a more detailed selection of attributes, the built regression models will be able adequately to predict a short period.

References

1. Tokunova, G.: Role of construction clusters in the development of the russian economy. World Appl. Sci. J. **23**(6), 812–816 (2013)
2. Stark, D.: The Sense of Dissonance: Accounts of Worth in Economic Life. Princeton University Press, Princeton (2009)
3. Russian cluster observatory. https://cluster.hse.ru/. Accessed 1 July 2018
4. Analytical Report—Fostering the international competitiveness of EU construction enterprises, European Construction Sector Observatory, https://ec.europa.eu/growth/sectors/construction/observatory_en, April 2017
5. Parygin, D., Sadovnikova, N., Kravets, A., Gnedkova, E.: Cognitive and ontological modeling for decision support in the tasks of the urban transportation system development management. In: IISA 2015—6th International Conference on Information, Intelligence, Systems and Applications, art. no. 7388073 (2016)
6. Arthur, W.B.: Increasing returns and path dependence in the economy [Text]/W.B. Arthur—Ann Arbor: University of Michigan Press (1994)
7. Grabher, G., Stark, D.: Organizing diversity: Evolutionary theory, network analysis, and postsocialist transformations. Restructuring networks: legacies, linkages, and localities in postsocialism. In: Grabher, G., Stark, D. (eds.) London and New York: Oxford University Press, pp. 1–32 (1997)
8. Porter, M.E.: On Competition. Harvard Business School Press, Boston, US, pp. 485 (1998)
9. Russia in Figures. 2017: Statistical Handbook/Rosstat - M., 2017 - 511 p http://www.gks.ru/free_doc/doc_2017/rusfig/rus17e.pdf
10. Anufriev, D., Petrova, I.Y., Shikulskaya, O.: Model of decision-making support in heterarchical system management of regional construction cluster. Commun. Comput. Inf. Sci. **754**, 317–330 (2017)
11. Rosstat official website. Construction. http://www.gks.ru/wps/wcm/connect/rosstat_main/rosstat/ru/statistics/enterprise/building/ Accessed 1 July 2018
12. Al-Gunaid, M.A., Shcherbakov, M.V., Skorobogatchenko, D.A., Kravets, A.G., Kamaev, V.A.: Forecasting energy consumption with the data reliability estimation in the management of hybrid energy system using fuzzy decision trees. In: IISA 2016—7th International Conference on Information, Intelligence, Systems and Applications, art. no. 7785413 (2016)

Actual Issues of Forensic-Environmental Expert Activity: Kazakhstan and International Experience

Kaliolla Seytenov

Abstract The perfection of forensic examination activities in the Republic of Kazakhstan in detecting and investigating of environmental crimes requires theoretical justification matching the modern level of development of criminalistics and general theory of forensic examination that includes systematization of forms of use of special knowledge and development of scientific and methodological support of specific types of forensic-environmental examination. The chapter describes the main problems and perspectives of the development of forensic-ecological activities in the Republic of Kazakhstan. To improve the effectiveness of forensic institutions, in the context of international experience in this area, it is proposed to unify the names of types of forensic environmental examinations, programs for additional professional education; to harmonize approaches to training forensic experts in environmental expert professions; to develop the quality management system, to carry out accreditation of forensic examinations institutions. Study of the international experience in the field of forensic environmental expert activity, joint development of its scientific and methodological support, implementation of promising approaches to the development of new types of forensic-environmental examination contribute to effective implementation of the integration function of international legal and human rights cooperation in this area.

Keywords Forensic-environmental examination · International experience

1 Introduction

In the administration of justice, a clear legislative regulation of proof is of special importance, which today is inconceivable without the use of special scientific knowledge, primarily through forensic examinations. Participation of an expert in

K. Seytenov (✉)
Institute of Forensic Expertise of the KAZGUU University,
Astana 020000, Kazakhstan
e-mail: ise.astana@yandex.ru

© Springer Nature Switzerland AG 2019
A. G. Kravets (ed.), *Big Data-driven World: Legislation Issues and Control Technologies*, Studies in Systems, Decision and Control 181,
https://doi.org/10.1007/978-3-030-01358-5_19

legal proceedings significantly expands the capacity of judicial and law enforcement bodies to establish the circumstances of the case and to adopt a procedural decision on their basis.

The expert's conclusion, drawn up by a person who is not interested in the outcome of the case, with the support of fundamental scientific provisions and approved research methods, is the source of factual data relevant to the case. At the same time, in the world practice more and more scientists are critical of the forensic examination and its role in the decision of accusatory or acquittal sentences [1].

Scientific developments, inconceivable for 20 years, not to mention 50 years ago, involve significant risks: the pace of change and the complexity of the methods used create certain difficulties for all participants in the proceedings, but, first of all, for those who do not have the necessary scientific knowledge [2]. That is why the standards of reliability and validity of forensic evidence used by the courts should be the standards that ultimately stem from the legal objectives of legal institutions, and not from the scientific purposes of scientific institutions [3].

The situation that has developed in Kazakhstan in the use of special knowledge in proving through forensic examinations now needs a drastic change in the regulatory legal framework regulating forensic activity.

To date, the country has been doing tremendous work on the legal reform of judicial and expert activities.

The new Law of the Republic of Kazakhstan "On Forensic Expert Activity" was adopted and put into effect on February 10, 2017. In order to improve the forensic work to the level of international accreditation standards, in October 2016 the World Bank project "Strengthening of Forensic Expertise" started. This project is implemented by the Ministry of Justice of the Republic of Kazakhstan and the Kazakh State Law University of the Ministry of Justice of the Republic of Kazakhstan with the support of Key Forensic Services, King's College London, QPA Strategies, Astana Garant Consulting.

The aim of the project is to study the international experience of organization of forensic examination activities, to develop institutional and legal mechanisms of strengthening of forensic examination in the Republic of Kazakhstan through:

- development of a system of effectiveness evaluation of forensic examination activities;
- development of a mechanism of a phased transition to a model of private forensic examination;
- development of regulatory documents required for implementation of forensic examination activities;
- organization of courses of qualification improvement of forensic experts in foreign universities and expert institutions.

The above conversions should be applied to such area as the forensic-environmental examination as well.

2 The Relevance of the Development of Forensic-Environmental Examination

What is the base of the relevance of the development of forensic-environmental examination in Kazakhstan?

First of all, the fact that along with the increase of consumption of natural resources, the intensive development of energy, industry, and agriculture in Kazakhstan, anthropogenic pressure on the environment is growing.

In the Message of the President of the Republic of Kazakhstan N.A. Nazarbayev to the people of Kazakhstan "Strategy "Kazakhstan-2050": a new political course of the established state", it is indicated that "… the most important part of the activities of the leadership of the Republic of Kazakhstan is to ensure environmental security, including protection of the environment and vital human interests from the possible negative impact of economic and other activities, emergency situations of natural and man-made character, liquidation of their consequences" [4].

The criminal legislation of the Republic of Kazakhstan provides for liability for environmental offenses and mandatory compensation for damage caused to the environment.

So, Chapter "Counteracting the Spread of Socially Dangerous Information on the Internet: A Comparative Legal Study" of the Criminal Code of the Republic of Kazakhstan, that entered into force on 1 January 2015, consists of 20 articles related to criminal environmental offenses. For example, the article 334 of the new Code provides for criminal liability for unauthorized use of mineral resources if these acts caused major damage. Criminal liability is also provided for violation of rules of protection of fish stocks while constructing of bridges, dams, realizing of explosive or other works, exploitation of water intake facilities or pumping mechanisms if it caused or could cause a mass death of fish or aquatic animals.

To determine causes, sources and consequences of negative impacts on the environment and compensation for environmental damage caused by illegal actions in the Forensic Examination Center of the Ministry of Justice of the Republic of Kazakhstan there has been initiated a research work on organization of a new complex direction of expert studies—forensic-environmental examination (FEE) [5].

To establish the actual circumstances of environmental crimes in the forensic examination institutions of the Republic of Kazakhstan forensic environmental and other forensic examinations are conducted. At the same time, there is a minimum number of forensic examinations in the above direction conducted in Kazakhstan, which does not allow to speak about serious counteraction to these threats and challenges.

The improvement of forensic expertise in the detection and investigation of environmental crimes is due to the needs of investigative and judicial practice, and it is impossible without a theoretical justification that corresponds to the current level of development of criminalistics and the general theory of forensic examination, including the systematization of the forms of the use of special knowledge, as well as scientifically substantiated determination of expert tasks, solved while conducting specific types of forensic-environmental examination or in the framework of the complex examination.

Forensic environmental expert activities in the Republic of Kazakhstan are the activities on organizing and carrying out of the forensic-environmental examination, its scientific-methodical and staff assistance with the use of special scientific knowledge in the field of ecology and related scientific, technical and economic sciences in the judicial proceedings.

The solution of environmental issues in most cases goes beyond the traditionally solvable forensic tasks and required a thorough study of the specialized literature and normative legal documents in the field of ecology and environmental protection. Increasingly, forensic examinations are carried out as complex expert examinations with the participation of experts from various expert fields using analytical equipment available in forensic institutions of the Republic of Kazakhstan, and in some cases involving well-educated persons with special knowledge in the field of ecology, related natural, technical and economic sciences. As a result of conducting complex examinations, the actual important circumstances of environmental offenses are being established.

The bulk of forensic research in Kazakhstan is conducted on the following environmental violations:

- pollution of soil by heavy metals, pesticides, oil and oil products (discharge and illegal dumping), destruction of nature of soil because of construction works, illegal dumping of waste and rubbish which leads to the loss of harvest or any other vegetation;
- illegal cutting of trees and shrubs in riverbeds and on hillsides that results in a change of land configuration, i.e. destruction of hillsides and alteration of riverbeds as their root system performs a drainage function, fixes hillsides and riverbeds, protects them against soil erosion and formation of landslides;
- the mass death of aquatic animals in ponds contaminated by industrial waste (as a result of accidental spills of oil products and discharges of industrial waste);
- illegal catching of fish with the use of illegal fishing methods, for example, electric current and fishing nets;
- illegal extraction of animals and plants included in the list of rare and endangered species of animals and plants included into the List of types of forensic examinations conducted in forensic examination bodies of the Ministry of Justice of the Republic of Kazakhstan [6].

It is also worth noting that the forensic ecological expertise is represented by one expert profession 19.1 "Forensic ecological research" and is included in the List of types of forensic examinations carried out by the forensic examination bodies of the Ministry of Justice of the Republic of Kazakhstan.

In forensic examination institutions of the system of the Ministry of Justice forensic-environmental examination includes 5 species. We share the view of Bekzhanov Zh.L., who proposed to expand the list of types of forensic-environmental examination conducted in the Republic of Kazakhstan up to 6 types: ecological-geological, ecological-hydrological, ecological-biological, ecological examination of objects of the urban environment, ecological-pedological and ecological-cost. Five of these species having different names are generally consistent with expert tasks similar to the types of expert tasks performed in Russia. The main difference is that in Kazakhstan forensic ecological-geological examination should be formed as a separate type. The proposal is due to the following. As you know Kazakhstan is actively carrying out activities of mineral production. Kazakhstan is one of the largest producers of oil, gas, uranium and other minerals. Violations of environmental law often happen while mining. That is why it is important to develop forensic ecological examination activities to establish actual circumstances of environmental offenses committed in the process of mining through conducting the ecological-geological examination.

3   The Perfection of the Regulation of Forensic Environmental Expert Activities

The perfection of the regulation of forensic environmental expert activities aimed at improving the quality of expert production and reducing the timing of forensic environmental assessments is impossible without changes in the field of professional additional education in expert environmental specialties.

The profile of the basic education of an expert should correspond to the tasks and research object that he faces. The person who appoints expertise and does not have knowledge in the field of ecology and related natural, technical and economic sciences is unable to accomplish this task; however, the competence of the forensic expert and the profile of his basic education are essential not only to assess the admissibility, but also the reliability of the conclusion.

In Kazakhstan, we have made significant progress in the training of forensic experts. In 2015–2016, the Institute of Forensic Expertise of the Kazakh State Law University of the Ministry of Justice of the Republic of Kazakhstan together with the Center of Forensic Expertise of the Ministry of Justice of the Republic of Kazakhstan introduced the specialty "Forensic expertise" in the classifier of professions of higher and postgraduate education of the Ministry of Education and Science of the Republic of Kazakhstan for the training of holders of a Master's degree and a degree of Doctor of Science.

In accordance with the order of the Committee for Control in the Sphere of Education and Science of the Ministry of Education and Science of the Republic of Kazakhstan of January 31, 2017, No. 111, the University was given a License for the right to conduct postgraduate educational activity in accordance with the master's degree in specialty 6M030500—"Forensic Expertise". One of the directions of specialized and scientific-pedagogical training in this specialty is "Forensic-ecological examination".

It should also be noted that the integration processes in the field of forensic environmental expertise have specific features. The basis of such integration is a uniform methodology of forensic examination, the joint development of which began in the 60s of the XX century.

One of the priority directions of the partnership of forensic examination institutions of the countries of near and far abroad is the achievement of high quality of forensic-environmental examination through the establishment of a quality management system conforming international standards. This area requires coordinated activities of accreditation of forensic examination institutions conducting forensic-environmental examinations. Implementation of these measures will allow effectively accomplish conducting complex multi-object forensic-environmental examination; improve the dynamics of development of scientific and methodological support for forensic-environmental examination; coordinate research; optimize information exchange; upgrade additional professional education; objectify assessment of opinions of experts; ensure the possibility of unimpeded use of conclusions of an expert-ecologist in the international courts.

Harmonization of the activities of forensic examination institutions of the Ministries of Justice of the member states of the EEU is of particular importance in connection with solving tasks aimed at formation of common external customs borders of its member countries; the implementation of a unified foreign economic policy and legal support for the functioning of the common market including in terms of membership in the World Trade Organization and functioning of the Court of the Eurasian Economic Union that should have the most complete list of competencies on use of modern scientific and technological achievements for the purpose of comprehensive and objective investigation of the circumstances that should be proved in the case under consideration.

4 Conclusion

The integration of each of the member states of the EEU in legal and economic space of the world community, as well as the increase of activities of all subjects of juridical proceedings related to the use of special knowledge in international courts, require, as the primary task, the accreditation of forensic organizations for confirmation of their technical competence in accordance with the international standards.

As a result of mutually beneficial cooperation among all countries in the field of forensic environmental examination activities it is expected to unify names and programs of additional professional education of environmental expert professions, to work together to develop and assess appropriateness of methodological support of forensic-environmental examination activities that will further allow them to be used as one of the preparatory stages of accreditation of forensic-examination institutions.

In general studying of experience in the field of forensic environmental examination activity, joint development of its scientific and methodological support, implementation of promising approaches to the development of new types of the forensic-environmental examination will contribute to effective implementation of integration function of international legal and human rights cooperation in the field of forensic environmental examination.

References

1. O'Brien, E., David, N., Black, S.: Science in the courts: pitfalls, challenges, and solutions. Philos. Trans. R. Soc. B. **370**(1674) (2015) https://doi.org/10.1098/rstb.2015.0062
2. The Rt Hon the Lord Thomas of Cwmgiedd: The Future of Forensic Science in Criminal Trials (2014). https://www.judiciary.gov.uk/wp-content/uploads/2014/10/kalisher-lecture-expert-evid ence-oct-14.pdf. Accessed 20 Apr 2015
3. Schauer, F.: Can bad science be good evidence? Neuroscience, lie detection and beyond. Cornell Law Rev. **95**(1191–1220), 12–14 (2010)
4. The Message of the President of the Republic of Kazakhstan N. A. Nazarbayev to the People of Kazakhstan "Strategy "Kazakhstan-2050": A New Political Course of the Established State" Astana 1 (2012). http://online.zakon.kz/Document/?doc_id=31305418&doc_id2=31312028#p os=1;-8&sub_id2=100&sel_link=1002707921. Accessed 1 Jul 2018
5. Bekzhanov, Z.L.: On the classification of forensic environmental examinations in the Republic of Kazakhstan. In: Materials of the International Scientific and Practical Conference "Problems of Classification of Forensic Expertise, Certification, and Validation of Methodological Support, Standardization of Forensic Expertise", pp. 27–28. Moscow (2016)
6. On the Approval of the List of Types of Forensic Examinations Carried out by the Forensic Examination Bodies of the Ministry of Justice of the Republic of Kazakhstan: Order No. 52 of the Minister of Justice of the Republic of Kazakhstan of January 26, 2015

Investment Management Technology with Discounting

Sergey L. Chernyshev

Abstract The actual task of realizing the attempt to combine the theory of management of the investment portfolio of Markowitz with the discount approach to the assessment of profitability is considered. The Markowitz portfolio theory, elements of mathematical statistics and economic theory, decision theory, the theory of risk management. Definitions have been found for the variance of the net present value of the portfolio of different projects. Its dispersion is found. The relationship of the net present value of the portfolio to the probability of risk is analyzed. Properties characteristic for the variance of an individual component of the net present value are also retained for the complete variance of the portfolio. The impact of the project timelines on the likelihood of risk is considered. The chapter shows that the average net present income is growing faster with an increase in the duration of projects than its variance, which leads to a decrease in the probability of risk. However, uneven changes in the timing of projects do not lead to a minimum probability of risk and the need for correction of investment shares.

Keywords Net present value · Portfolio · Discounting · Dispersion
Diversification · Optimal shares · The probability of risk

1 Introduction

The technology of investment management is based on minimizing the probability of the investment risk. This minimization is ensured by the distribution of investments between different projects, on which the diversification method is based. Markowitz et al. [1–3] and other authors [4–6] have developed a theory of portfolio management of production projects. Reducing the likelihood of a project portfolio risk is possible in two ways: maximizing the average income of the portfolio, limiting the dispersion of this income and minimizing the variance of income with a restriction on the average

S. L. Chernyshev (✉)
Bauman Moscow State Technical University, Moscow 105005, Russian Federation
e-mail: chernshv@bmstu.ru

© Springer Nature Switzerland AG 2019
A. G. Kravets (ed.), *Big Data-driven World: Legislation Issues and Control Technologies*, Studies in Systems, Decision and Control 181,
https://doi.org/10.1007/978-3-030-01358-5_20

income. The first method is difficult to implement in practice since it is necessary to select more profitable projects in the portfolio [7]. Therefore, the second method has become more widespread. The management technology described in this chapter is based on this method. However, the usual approach to managing a portfolio of production projects that are of a long-term nature must consider the times at which payments are made in the projects that make up the portfolio [8]. This feature of production projects makes it necessary to use the discount method [9]. In this method, four indicators of economic efficiency are considered [10]: the net present value *NPV* (it is equal to the difference between the discounted amount of sales proceeds and the discounted amount of costs), the internal rate of return *IRR* (the estimated interest rate at which *NPV* = 0), as well as an additional discount payback period *PP* and profitability index *PI*. This chapter is devoted to the technology of investment portfolio management adjusted for the long-term production projects based on the discount approach.

2 Net Present Value of the Projects Portfolio and Its Dispersion

If a portfolio has N projects then net present value *NPV* of the portfolio will be determined from the following expression:

$$NPV = \overline{NPV} + \sum_{k=0}^{T} \sum_{r=1}^{N} S_r(k)\varepsilon_{Rr}(k)(1+i)^{-k}, \tag{1}$$

where is the average value of net present value,—natural sales volumes,—random variation in prices, i—discount rate, T—time period under review.

Random variation of the net present value is determined by the addend in (1).

In accordance with (1) net present value of the portfolio is a random variable. Then its dispersion will be determined by the expression [9]

$$\sigma_{NPV}^2 = \sum_{r=1}^{N} \left\{ \sigma_{Rr}^2 S_{rd}^2 + \sum_{\substack{m=1 \\ m \neq n}}^{N} \sum_{n=1}^{N} r_{mn} \sigma_{Rm} \sigma_{Rn} S_{mnd}^2 \right\}, \tag{2}$$

where, $S_{rd}^2 = \sum_{k=0}^{T} S_r^2(k)(1+i)^{-2k}$, r_{mm} is a correlation coefficient between prices of products of different projects, σ_{Rm} and σ_{Rn} are their standard deviations,

Management to reduce the likelihood of a portfolio risk is to minimize variance (2) by selecting the appropriate projects in the portfolio and the proportion of their financing, that is, through diversification.

3 Share Management

Optimal shares c_{ropt} by Markowitz are defined by a criterion of dispersion minimum of the net present value of the portfolio. In our case shares shall be determined as

$$c_r = \frac{S_r}{S_\Sigma}$$

the share of production of r project in the total value of production $S_\Sigma = \sum_{n=1}^{N} S_n$ of all projects.

Let's consider for simplicity the special case when in all projects constant sales are being implemented by an ordinary annuity, that is $S_r(k)\bar{R}_r(k) = const$, and also when these projects are not correlated among themselves. For this case formula (2) including share, financing takes the form

$$\sigma_{NPV}^2 = \sum_{r=1}^{N} \sigma_{Rr}^2 S_r^2 d_{Ti},$$

where $d_{Ti} = \frac{1-(1+i)^{-2T}}{(1+i)^2-1}$ is a coefficient of reduction of variance. Average net present value for this case will be determined as

$$\overline{NPV} = \sum_{r=1}^{N} (S_r \bar{R}_r a_{Ti} - K_r), \ NPV_{min} = \overline{NPV} - 2\sigma_{NPV}2$$

where $a_{Ti} = \frac{1-(1+i)^{-T}}{i}$ is a coefficient of reduction of a constant rent ordinary annuity, are one-time investments.

Analysis of these expressions showed that, if there is a rise in prices, a speed of growth of $\overline{NPV}$ increases with the increase of projects timings compared to growth of σ_{NPV} that leads to a decrease of risk probability. It is typical for the case when the timings of projects in the portfolio are the same.

However, the technology of project portfolio management is changed if the timings of the projects change. In this case

$$\sigma_{NPV}^2 = \sum_{k=0}^{T_r} \sigma_k^2 = \sum_{r=1}^{N} \left\{ \sigma_{Rr}^2 \sum_{k=0}^{T_r} S_r^2(k)(1+i)^{-2k} \right\}$$
$$+ \sum_{m=1}^{N} \sum_{\substack{m=1 \\ n\neq 1}}^{N} \left\{ r_{mn} \sigma_{Rm} \sigma_{Rn} \sum_{k=0}^{T_r} S_m(k)S_n(k)(1+i)^{-2k} \right\}$$

where T_r is the period of r project. Let's consider for simplicity a portfolio of two projects ($N = 2$). In this case, in accordance with the known approach [2], the

Fig. 1 Dependence of the optimal share of the discount rate

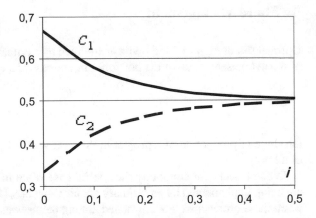

minimum probability of risk corresponds to the optimal share of investment in the projects. These optimal shares for uncorrelated prices are calculated as

$$c_1 = \frac{\sigma_2^2 d_{T_2 i}}{\sigma_1^2 d_{T_1 i} + \sigma_2^2 d_{T_2 i}} \text{ and } c_2 = \frac{\sigma_1^2 d_{T_1 i}}{\sigma_1^2 d_{T_1 i} + \sigma_2^2 d_{T_2 i}} \cdot c_2$$

It is obvious that with the change in the timing of projects optimal shares change as well. Therefore, investment management technology will depend both on the timing of projects, and interest discount rates i. Figure 1 shows the dependence of the optimal shares of investment projects on the discount interest rate, which, as is known, depends on the level of inflation.

So, with the increase of the interest rate of inflation the discount rate should also rise. And in these conditions, optimal shares should be managed to provide the minimum probability of risk. For the dependence described by the Fig. 1 there are the following projects timings $T_1 = 5$ years, and $T_1 = 10$ years. It follows from the Fig. 1, that the management of investment projects with different terms of implementation depends on the interest rate and the management technology should take into account this dependence.

It is also important to consider how the net present value of the portfolio changes with this shared management. If the flows of payments in projects are permanent ordinary annuity rents with different terms and lump-sum investments, then each project will have a net present value, which is defined as

$$\overline{NPV} = \sum_{r=1}^{N} (S_r \bar{R}_r a_{T_r i} - K_r)$$

The optimal releases under the above share management will be determined as

$$S_{ropt} = c_{ropt} S_{\Sigma}$$

Fig. 2 The change in the average net present value with increasing discount rates (1—with share management, 2—without it)

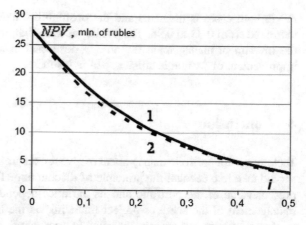

Fig. 3 The change in the lower threshold of the net present value for a risk probability of 5% with an increase in the discount rate (1—with share management, 2—without it)

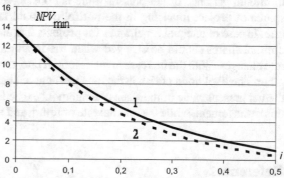

Figure 2 shows the change in the average value of $\overline{NPV}$ while changing the discount rate.

It expectedly decreases with its increase, for example at the rate of, the level of inflation according to estimates is more than 37% that significantly reduces the modern value of the project. However, as it can be seen from this figure when managing shares (curve 1) the value $\overline{NPV}$ is slightly higher than without the share management (curve 2). While calculating, a special case was considered, when $S_\Sigma \bar{R}_1 = 5$ million rubles, $S_\Sigma \bar{R}_2 = 10$ million rubles, $T_1 = 5$ years, $T_2 = 10$ years, $K_1 = 2$ million rubles, $K_2 = 4$ million rubles.

Figure 3 shows the dependence of the minimum threshold of net present value for the probability of risk equal to 5% with share management and without it $(NPV_{min} = \overline{NPV} - 2\sigma_{NPV})$. You can see that with share management this threshold is higher, i.e. the probability of risk of 5% is provided together with a larger profitability of the portfolio. If you set this lower threshold (the same for both cases), the probability of risk with share management will likely be lower than without it.

Calculation of internal rate of return showed that share management in this example, though not significantly, but increased from 80 to 82%. Discount payback period

PP in both cases is the same and the profitability index with share management increased from 0.83 to 0.96. These data indicate that, in addition to risk reduction, the investment management technology described in this chapter can lead to the improvement of economic efficiency of investment.

4 Conclusion

Technology for portfolio management of investment projects based on diversification should take into account the principle of discounting. The average value of the net present value of the portfolio and its variance depend on the discount rate. The management of the share of project financing on the basis of the criterion of the minimum variance of net present value is considered. It is shown that the optimal shares of project financing in the portfolio should take into account the discount rate. In case of unequal timeframes for projects that are included in the portfolio, the optimal shares should be changed when the inflation rate included in the discount rate changes. With the increase in the discount rate, the optimal shares approach each other. The likelihood of risk in the management of shares is reduced. It is shown that the net present value of the share management exceeds the amount obtained without this management, while the internal rate of return and the profitability index increase.

References

1. Gupta, F., Markowitz, H.M.: Theory of portfolio selection. In: Institutional Investment Management: Equity and Bond Portfolio Strategies and Applications, pp. 13–40 (2012)
2. Markowitz, H.M.: Harry Markowitz: Selected Works. World Scientific-Nobel Laureate Series, vol. 1, p. 716. World Scientific, Hackensack, New Jersey (2009)
3. Markowitz, H.M.: Portfolio Selection: Efficient Diversification of Investments. Wiley, New York (reprinted by Yale University Press, 1970, ISBN 978-0-300-01372-6; 2nd ed. Basil Blackwell) (1991)
4. Rajagopal, S., McGuinn, P., Waller, J.: Project Portfolio Management: Leading the Corporate Vision. Palgrave Macmillan, Basingstoke (2007)
5. Sanwal, A.: Optimizing Corporate Portfolio Management: Aligning Investment Proposals with Organizational Strategy, p. 224. Wiley, NJ, USA (2007)
6. EPMC Inc., Stratton, M.J., Wybraniec, M., Tekumalla, S., Mark Stabler, M.: Project Portfolio Management: A View from the Management Trenches, p. 256. Wiley, NJ, USA (2009)
7. Skaf, Mazen A.: Portfolio management in an upstream oil and gas organization. Interfaces **29**(6), 84–104 (1999)
8. Shcherbakov, M., Groumpos, P.P., Kravets, A.: A method and IR4I index indicating the readiness of business processes for data science solutions. Commun. Comput. Inf. Sci. **754**, 21–34 (2017)
9. Chetyrkin, E.M.: Financial Mathematics, p. 392. Publishing House "Delo" of the Russian Academy of National Economy and Public Service, Russia, China (2011)
10. Shcherbakov, S.L.: Risks of the portfolio of production projects taking into account the random scatter of market prices and discounting. Mechanical Engineering and Computer Technologies **11**, 153–161 (2016)

Development of Communication as a Tool for Ensuring National Security in Data-Driven World (Russian Far North Case-Study)

Igor Gurlev, Elena Yemelyanova and Tatiana Kilmashkina

Abstract The issues of improving communication services in the Far North of Russia at the present stage are extremely relevant in connection with the plans for the development of the Northern Sea Route and its transformation into a global transit sea route. The Far North with its huge reserves of hydrocarbons, other minerals, and bioresources is of strategic importance in such areas of national security as economic, social, environmental, etc. Therefore, the issues of providing this part of the country with various communication services are also of strategic importance. The chapter presents data on cargo turnover along the Northern Sea Route, the state, and development of the nuclear icebreaking fleet; shows the dynamics of natural gas production in Russia over the past decade and the forecast of gas production for the future up to 2035. The peculiarities of the development of various types of communication are analyzed taking into account the peculiarities of the severe climate of the region: radio relay and fiber-optic communication lines, as well as satellite communications provided by the space vehicles located in the geostationary orbit and in highly elliptical orbits. The solution of the complex task of providing all consumers of this high-latitude region with high-quality high-speed information and telecommunication services contributes to the smooth year-round operation of the infrastructure management system of the Northern Sea Route.

Keywords Far North · Northern Sea Route · Nuclear icebreaking fleet
Various types of communication · Radio relay communication
Fiber-optic communication lines · Satellite communication

I. Gurlev (✉) · E. Yemelyanova · T. Kilmashkina
Academy of Management of the Ministry of Internal Affairs of Russia, Moscow 125993, Russian Federation
e-mail: gurleff@mail.ru

E. Yemelyanova
e-mail: eev-rusinovo@yandex.ru

T. Kilmashkina
e-mail: kilmashkinnf@yandex.ru

© Springer Nature Switzerland AG 2019
A. G. Kravets (ed.), *Big Data-driven World: Legislation Issues and Control Technologies*, Studies in Systems, Decision and Control 181,
https://doi.org/10.1007/978-3-030-01358-5_21

The control system for uninterrupted year-round operation of the infrastructure of
the Northern Sea Route

1 Introduction

The Far North of Russia, where the world's largest reserves of solid minerals, hydrocarbons, and bioresources are located, is a strategic area of vital importance for the
Russian economy. In addition, there are some internal and external factors of a political and geostrategic nature that determine the uniqueness of the Far North and make
its development an important national priority:

– firstly, this is the ensuring of national security in the economic sphere in the Arctic
  direction, maintaining the ecological balance of fragile northern natural systems
  and providing the social and economic well-being of the indigenous population of
  these territories;
– secondly, it is the effective and rational use of the natural resources of the Arctic
  region: hydrocarbons, solid minerals and biological resources;
– thirdly, it is the provision of transport and information security of the vast and
  sparsely populated northern regions of Russia, etc.

The Far North of Russia is one of the main oil and gas bearing regions of the world
[1]. According to expert estimates, about 100–120 billion tons of hydrocarbons in
oil equivalent are concentrated in its bowels, 80% of which is gas. According to
forecasts of specialists for the next 30 years, global energy use will increase by two
thirds, and by 70% it will be covered by hydrocarbons [2].

The data of the dynamics of gas production in Russia over the past decade, (according to the Statistical Yearbook of the World Energy Sector, 2017) are presented in
Table 1.

Many states of Western Europe and the Asia-Pacific region link their energy needs
with Russia's gas resources, so the development of gas production in Russia is of
global geostrategic, economic and political importance.

The development of the Russian gas industry for the near and medium term is
determined by the adjusted Russian Energy Strategy for the period up to 2035.
According to it, by 2035 it is planned to increase gas production to 821–885 billion
m^3 per year [3], so natural resources further will keep their importance for Russian
state development [4] and its regions [5].

Already, almost 30% of the gas produced is exported from the country, and in
2020–2035s its export should increase even more significantly. Therefore, in order
to achieve the goals set by the Energy Strategy, in the coming years, new centers for
the extraction of "blue fuel" should be created in Russia. Such centers, for example,
can become open numerous gas fields located in the Yamal Peninsula.

The public joint-stock company (PJSC) "Novatek" is building a liquefied natural
gas (LNG) plant with a total capacity of 16.5 million tons in the port of Sabetta in
the Yamal peninsula within the framework of the Yamal-LNG project on the basis of

Table 1 Gas production in Russia (2006–2016)

Years	2006	2007	2008	2009	2010	2011	2012	2013	2014	2015	2016
Annual gas production	640	635	651	583	657	673	658	685	630	624	628

the large Yuzhno-Tambeyskoye gas field. Three lines will be constructed, each with a capacity of 5.5 million tons of LNG per year. According to the Minister of Energy Alexander Novak, the first LNG transit operations will start at November 2018 [6].

In recent years, the priority development of the Far North and the return of Russia to the Arctic has become an urgent economic and political task. One of the most important tasks is the development of the Northern Sea Route.

The length of the sea route through the Suez Canal between the ports of Rotterdam and Vladivostok is 12,840 nautical miles, while the length of the sea route along the northern coast of Russia if counted from Rotterdam to Vladivostok, is shorter by more than 2 times and is 5770 nautical miles.

The Federal Law No. 132-FL of 28.07.2012, "On Amending Certain Legislative Acts of the Russian Federation Regarding State Regulation of Merchant Shipping in the Water Area of the Northern Sea Route", establishes the official water area of the Northern Sea Route.

The Northern Sea Route along the northern coast of Russia, in comparison with the southern route through the Suez Canal, provides:

– a significant reduction in the duration of the voyage;
– a significant reduction in fuel costs;
– a reduction of staff costs;
– a reduction of the cost of payment for the freight of the vessel;
– a complete absence of the threat of the pirates attack;
– no restrictions on the dimensions of the vessel;
– no payment for the passage of the vessel along the Northern Sea Route.

At the same time, there is payment for icebreaker support.

The main ports of the northern route are Murmansk, Arkhangelsk, Naryan-Mar, Sabetta, Yar-Sale, Tazovsky, Dixon, Igarka, Khatanga, Tiksi, Green Point, Pevek, Cape Schmidt, Providence Bay, Petropavlovsk-Kamchatsky, Vladivostok.

The main Russian users of transportation services along the Northern Sea Route are Gazprom, Norilsk Nickel, Lukoil, Rosneft, Novatek and other large companies.

2   Russian Far North Infrastructure, Strategic Resources, and Cargo Flows

The data collected from the different sources for cargo transportation along the Northern Sea Route are given in the Table 2.

As it can be seen from Table 2, at the present time, the basis of cargo flows along the Northern Sea Route is internal cargo transportation to meet the needs of settlements and enterprises in the Far North. Transit freight flow is still substantially insignificant and quite unstable.

Currently, four nuclear and four diesel-electric icebreakers operate along the Northern Sea Route and the mouths of the northern rivers.

Table 2 Cargo traffic along the Northern Sea Route

Years	2010	2011	2012	2013	2014	2015	2016
Total cargo traffic (million tons)	2.90	3.11	3.60	3.93	3.98	5.39	6.92
Transit cargo traffic (million tons)	0.11	0.83	1.25	1.12	0.27	0.04	0.21

Built for the implementation of the Yamal-LNG project, the first of a series of 15 ice-breaking tankers, a "Christophe de Margerie" tanker, owned by PJSC "Sovcomflot", performed a test flight along the Northern Sea Route from August 11 to 17, 2017 from Cape Zhelaniya (Cape of Wish) on the Novaya Zemlya archipelago (New Land archipelago) to Cape Dezhnev in Chukotka for 6.5 days without icebreaker escort.

For today, the average time for icebreaking steering of vessels from Murmansk to the Bay of Providence during navigation is 10.6 days.

Navigation along the Northern Sea Route is usually carried out within 3–5 months of the year, depending on the ice conditions.

In order to substantially increase the navigation time along the Northern Sea Route, the new nuclear icebreakers are being built in the Russian Federation: nuclear icebreakers of the "Arctic" series of Project No. 22220 (LK-60YA) with a capacity of 60 MW (the hull of the "Arctic" nuclear icebreaker was launched on June 16, 2016), the nuclear icebreakers of the "Siberia" series (the hull of the "Siberia" nuclear icebreaker was launched on June 8, 2017) and the nuclear icebreakers of the "Urals" series.

On July 1, 2017, during the International Maritime Defense Show in St. Petersburg, the Deputy Prime Minister Dmitry Rogozin specified that the nuclear icebreakers "Arctic", "Siberia" and "Ural" will be handed over to the customer in 2019, 2020 and 2021 respectively.

In order to ensure uninterrupted year-round navigation along the whole route of the Northern Sea Route, a super-powerful icebreaker of the "Leader" series of Project 10510 (LK-110Ya, LK-120Ya) with a capacity of 110–120 MW is currently being designed in Russia. To solve this problem, it is necessary to build and put in commission three icebreakers of the "Leader" series that will be able to make their way through ice fields with a thickness of 4.3 m and lay a navigable canal up to 50 m wide for the passage of a caravan of large-capacity vessels at a speed of 5–10 knots. With an ice thickness of 3.5 m, the average speed will be 15 knots or more. Construction of the first icebreaker of the "Leader" is planned no later than 2030.

To reduce the length of the route, in relation to the projected super-power nuclear icebreakers of the "Leader" series, a new high-latitude international transit route is currently being developed, north of Novaya Zemlya, Northern Earth Islands, Novosibirsk Islands and Wrangel Island.

3 The Strategic Task of Communication Infrastructure Modernization

However, the successful and effective solution of tasks for the development of the Far North and the transformation of the Northern Sea Route into a global transport structure is impossible without a modern communication infrastructure.

The creation of an infrastructure that provides reliable, high-quality and uninterrupted communications as an important part of the management system for the development of the Far North of Russia should be considered a separate strategic task.

Thus, the intensive development of the regions of the Far North requires serious investments not only in the creation of transport, production, and technological facilities but also in building a high-quality and reliable telecommunications component.

The choice of technology for building the communication infrastructure is largely predetermined by the natural features of the Far North: an unstable geomagnetic situation in high latitudes, severe climatic conditions, the presence of many water barriers, permafrost, significant distances of production sites from each other and from the parent enterprise, exploration technology, operating deposits, transportation of the final product and other conditions. For example, after carrying out exploration work or stopping the exploitation of the field, the equipment must be moved to another location and the laying of communication cables in such cases is impractical, and the choice of a wireless type of communications is a faster, simple and economically feasible solution.

Considering the above and other factors in the Far North, three technologies for building the communication infrastructure were most widely used:

- radio relay communication;
- fiber-optic communication;
- space communication.

3.1 Radio-Relay Communication

One of the most economical and fast ways to organize information flows over long distances is the radio-relay linkage of the line of direct visibility. And now the analog trunk communication lines are replaced by modern digital radio relay stations (DRRS), which have high capacity and high-quality communication [7].

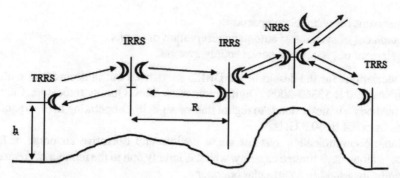

Fig. 1 RRL route of different nature

For the DRRS of main and intra-zone lines, there is a remote maintenance system that programmatically supports the level of management of network elements and the network as a whole, as well as provides control, management, and maintenance of equipment. Such stations operate, as a rule, in the frequency range 3.4–11.7 GHz; their throughput is 155 Mbit/s and more and signaling is carried out using multi-position types of modulation.

Radio-relay lines of direct visibility easily cross the hard-to-reach terrain of the Far North. in the regions of permafrost, where it is not always economically feasible to lay cable trunks. In addition, when creating branched networks, radio relay links are most effective.

To provide line-of-sight radio communications, the antennas are mounted on masts at a certain height. The heights of the masts depend on the length and geographical profile of each interval between neighboring stations and can reach 120 m. The construction of higher antenna towers is economically unprofitable.

The radio-relay communication line consists of terminal, intermediate and node radio relay stations (RRS), which carry out serial multiple retransmission (reception, conversion, amplification, transmission) of radio signals.

Theoretically, the intermediate stations of the radio relay link (RRL) can be installed at a distance of 40 km from each other or more, but in practice the masts with radio-relay communication antennas are set at a distance of 20–30 km from each other, for example, because of the climatic conditions or when there are significant obstacles along the RRL route of different nature, etc. (Fig. 1).

Radio relay link (TRRS—terminal radio relay station, IRRS—intermediate radio relay station, NRRS—node radio relay station, R—the distance between RRS, h—height of antenna installation on the mast).

In frequency bands above 11 GHz, the span length between adjacent RRS is inversely proportional to the frequency.

The structure of any radio relay station includes the following equipment:

- antenna-feeder devices;
- transceiver equipment;
- tele-control, tele-signaling, and service communication equipment;

– equipment for channel compaction;
– equipment of systems of automatic reservation of trunks;
– equipment of guaranteed power supply systems.

Calculations for the design of the DRRS are carried out in accordance with the State Standard R 53363-2009 "Digital radio relay lines. Quality Indicators. Calculation methods", which extends to digital line-of-sight lines operating in the frequency bands from 3.4 to 40.5 GHz.

Radio communication systems allow flexible and operative coverage of large areas, but they have limited capacity, which is largely due to the number of frequency assignments issued to a particular operator.

Promising areas for the development of radio relay communication are:

– use of packet transmission in a radio frame;
– use of more complex quadrature amplitude modulation schemes (QAM-512, QAM-1024);
– use of higher carrier frequencies in the 70–80 GHz bands.

The use of the 70–80 GHz band allows the RRS equipment to apply 100 MHz bandwidth available for transmission, which in turn allows achieving 1 Gbit/s throughput. A significant drawback of this range is strong attenuation at atmospheric precipitation, limiting the range of communication, depending on climatic factors to 2.5–5 km with an unavailability factor of 0.001%.

In the communication networks of gas producing enterprises, for example, members of the PJSC Gazprom Group, the share of radio relay communication is currently more than 60%. Such predominance of radio relay communication over fiber-optic and satellite communication technologies is due to:

– the history of the development of communication technology;
– a large, sparsely populated and little-developed territory of the Far North;
– features of the landscape of the northern territories;
– large distances between populated areas and deposits, fisheries, compressor stations, etc.

Radio-relay linkage of the line of sight is more stable than distant tropospheric communication, but just like tropospheric and satellite communications, depends on the state of the atmosphere.

Another important feature of radio relay communication is the fact that the load of the radio frequency range up to 11 GHz is currently such that the means of radio communication itself are forced to operate in the combined frequency bands, in which radar, radio navigation, and radio telemetry facilities also work. There is a problem of electromagnetic compatibility of various radio facilities, which requires a solution not only at the national but also at the international level. The electrical characteristics of communication systems that determine distortions of transmitted information are identified in the domestic lines of Russia by the standards of the Unified Automated Communication Network (UACN), in international lines—by the recommendations of the International Telecommunication Union (ITU).

3.2 Fiber-Optic Communication

One of the main types of communication, which does not depend on the severe climatic conditions of the Far North, is a connection based on the use of the properties of the optical fiber.

To provide users, trunk, zonal, field, local, interstation connecting, distribution and subscriber fiber-optic communication lines (FOCL) are designed and built.

In addition to numerous technological communication lines along the gas pipelines and oil pipelines in Yamal, optical communication lines of commercial telecommunication companies are laid.

In the years 2000–2014, PJSC "Rostelecom" constructed a trunk line "Northern Optical Stream" from Yekaterinburg through Nyagan, Khanty-Mansiysk, Surgut, Noyabrsk, NovyUrengoy to Salekhard with a length of 3500 km. The total length of fiber-optic communication lines functioning in the system of the "Northern Optical Stream" is today more than 15,000 km.

Other telecom operators are also actively developing the backbone infrastructure in the Far North, in many respects repeating the route of the "Northern Optical Stream".

Closed Joint-Stock Company (CJSC) "Trans-Tele-Com-Company" has the necessary backbone infrastructure in almost all cities covered by the "Northern Optical Stream". In September 2016, CJSC "Trans-Tele-Com-Company" completed the fiber optic link to the city of Labytnangi, located near Salekhard.

Public Joint-Stock Company (PJSC) "Mobile-Tele-Systems" over the past two years has increased the total length of fiber optic links in the Khanty-Mansi and Yamal-Nenets Autonomous Districts by more than 2000 km. The optical highway of PJSC "Mobile-Tele-Systems" connected the cities of Noyabrsk, Muravlenko, Gubkinsky, Novy Urengoy, Pangody, and Nadym, as well as Surgut and Nizhnevartovsk, Novy Urengoy and Yamburg [8].

On September 23, 2017, Norilsk, Kayerkan and Dudinka held a celebration on the arrival of the Taimyr broadband Internet. More than 950 km of fiber optic cable from Novy Urengoy to Norilsk were laid at the expense of PJSC "Nornikel". The route runs through the territory of the Yamal-Nenets Autonomous District and the Krasnoyarsk Territory. The bandwidth of this fiber optic link is 40 Gbit/s [9].

3.3 Space Communication

Satellite communication systems are actively developing in the Far North. These systems have several significant advantages in comparison with cable networks. They have a much wider coverage of the territory because do not depend on the infrastructure of land communications, and therefore in remote and sparsely populated regions of Russia are the most optimal technical and economic solution. In addition, the presence of a satellite communication system makes it possible to provide com-

Table 3 Frequency bands for satellite communication systems

No.	Name	Frequency band (GHz)	Diameter of antenna (m)	Application area
1	2	3	4	5
1	L-range	1452–1550 and 1610–1710		Mobile satellite communication
2	S-range	193–270		Mobile satellite communication
3	C-range	340–525 and 5725–7075	2.4–2.5	Fixed satellite communication
4	Ku-range	1070–1275 and 1275–1480	0.6–1.5	Fixed satellite communication, satellite broadcasting
5	Ka-range	1540–2650 и 2700–3020	0.3–0.9	Fixed satellite communication, inter-satellite communication

munication for sea vessels, numerous nomadic reindeer breeders, geological parties, as well as to combine internal communications of the branches of extractive companies, populated areas, state institutions and law enforcement units of the Ministry of Internal Affairs into a single information network [10].

Satellite communication is today a widely used technology, through which telephone and facsimile communications are provided, Internet access, video conferencing and the like.

For the construction of communication networks, large territorially-distributed companies currently widely use satellite systems of VSAT technology (Very Small Aperture Terminal). One of the most significant advantages of this technology is its possible complete independence from the availability of local terrestrial Internet providers. To communicate using VSAT technology, only electricity and direct visibility to the satellite are needed [11].

Unlike terrestrial wired communication networks, which are subject to such hazards as the failure of network equipment, failures and accidents in electrical equipment networks, including breakage and damage to the cable for various reasons (for example, rodents or the movement of frozen grounds), satellite communication is spared from all these shortcomings.

In accordance with the ITU Regulations for Satellite Communication Systems in Russia, several frequency bands have been allocated (Table 3) [12].

The telecommunication and information resource of the Russian market of space communication technology VSAT is almost entirely provided by geostationary satellites of two domestic companies: the Federal Space Agency "Space Communications" (satellites of the "Express-AM" series) and "Gazprom Space Systems" ("Yamal"-type satellites) [13].

Russian satellite communication and broadcasting systems operate mainly in the C and Ku-bands. In recent years there has been a transition of satellite communication technology VSAT to a higher-frequency Ka-band, in which the antennas are smaller in size [14].

With the help of communication satellites located in geostationary orbits, it is possible to quickly form a network infrastructure with the high-reliability indicators; therefore VSAT satellite channels are widely used in the construction of distributed corporate and state networks. For such channels, a sufficiently high level of encryption and data protection is provided. VSAT communication systems protect the transmitted information much more reliable than other communication technologies and are selected by users to reserve available channels, as they are obviously safer from a technical point of view and are maximally protected from damage and failures.

The largest providers of satellite communications technology VSAT in the Russian market are the companies: "Altegro Sky", "Web Media Services", "IP Net", "Amtel Communications" and others [15].

To ensure a stable connection above the 76th parallel north latitude it is necessary to create satellite communication and broadcasting systems using highly elliptical orbits (HEO) satellites. Such satellites will be able to provide communication and broadcasting services to the Arctic zone, the Far North, and other Russian territories. According to the Minister of Communications and Mass Media Nikolai Nikiforov, in 2015, proposals for the creation of such a communication system were sent to the Federal Space Agency to be included in the draft of the Russian Federal Space Program for 2016–2025 years [16].

4 Conclusion

In conclusion, it should be noted that the development of the Northern Sea Route promotes the diversification of Russia's economy: improvement of the technologies of production and delivery systems of hydrocarbons; development of the icebreaking ship-building; modification of the marine nuclear power plants, propulsion engineering; radio relay communication systems; fiber-optic and satellite communications; manufacture of satellites and their component base etc.

At the same time, the characteristics of Russia as a "gas station" or "a lighter" are obviously insincere, since the presence of large reserves of oil, gas and other minerals is a strategic advantage in the economic, energy and other spheres of ensuring Russia's national security.

References

1. Mareš, M., Laryš, M.: Oil and natural gas in Russia's eastern energy strategy: dream or reality? Energy Policy **50**, 436–448 (2012) (November)

2. Namsaraev, Z.B., Gotovtsev, P.M., Komova, A.V., Vasilov, R.G.: Current status and potential of bioenergy in the Russian Federation. Renew. Sustain. Energy Rev. **81**(Part 1), 625–634 (2018)
3. The energetic strategy of Russia for the period up to 2035. http://www.energystrategy.ru/ab_i ns/source/ES-2035_09_2015.pdf. Accessed 1 July 2018
4. Studin, I.: Russia: Strategy, Policy and Administration, 412 p. Palgrave Macmillan, UK (2018)
5. Bondarenko, Y.V., Azarnova, T.V., Kashirina, I.L., Goroshko, I.V.: Mathematical models and methods of assisting state subsidy distribution at the regional level. In: International Conference "Applied Mathematics, Computational Science and Mechanics: Current Problems", 18–20 Dec 2017. Voronezh, Russian Federation
6. Russian LNG will go through Norway. Neft Kapital, 2018. http://www.oilcapital.ru/news. Accessed 29 June 2018
7. Bykhovsky, M.A., et al.: Fundamentals of Design of Digital Radio Relay Communication Lines, 332 p. Hotline Telecom, Moscow (2014)
8. "Norilsk Nickel" brought FOL to the North. https://www.comnews.ru/content/109734/2017-0 9-26/nornikel-privel-vols-na-sever. Accessed 1 July 2018
9. Rostelecom completes the construction of Nord Stream. https://www.comnews.ru/node/8171 3#ixzz4O6Wkldcr. Accessed 1 July 2018
10. Vysotsky, G.: VSAT network services and their consumers. Tele-satellite (3), 20–28 (2011)
11. Gurlev, I.V.: Methods and methods for ensuring the security of information transmitted over the satellite network of VSAT technology. Internet J. "Naukovedenie" **9**(3). http://naukovede nie.ru/PDF/85EVN317.pdf. Accessed 1 July 2018 (2017)
12. Kolyubakin, V.: What is VSAT. Tele-Sputnik (7), 6–8 (2015)
13. Maltsev, G.N.: Network information technologies in modern satellite communication systems. Inf. Control Syst. (1), 33–39 (2007)
14. Dodel, H., Eberle, S.: Satellitenkommunikation, p. 466. Springer, Berlin (2007)
15. Satellite operators of Russia and CIS: rating of Russian providers. http://www.kp.ru/guide/sp utnikovye-operatory.html. Accessed 1 July 2018
16. Four satellites can form an Arctic grouping of Russia. http://minsvyaz.ru/ru/events/33024. Accessed 1 July 2018

Analysis of the Data Used at Oppugnancy of Crimes in the Oil and Gas Industry

Dmitry Vasilev, Evgeny Kravets, Yuriy Naumov, Elena Bulgakova
and Vladimir Bulgakov

Abstract The chapter dwells on criminological characteristics of economic crime, related to theft in fuel and energy industry. The purpose of the chapter is to analyze collected data, crime dynamics in order to curb new high tech violations in strategically important branches, including, of course, the fuel and energy complex. It is exactly due to the above circumstances that the chapter was written. This research is also based on methods of system analysis and synthesis. The methods of scientific cognition used, such as analysis of statistical data, contributed to better understanding of the social danger of crime against property in fuel and energy industry and to the identification of basic determinants of the crime under review. The result of the work is general theoretical substantiation of using large data technology to fight this type of crime. The application sphere of the work results is not limited to theoretical research and can be used in the practice of law enforcing and judicial bodies. By classification, theft in the fuel and energy complex is rated as economic crime, which can be fought and prevented by applying large data technology, including data, based on coordination between oil transporting companies and law enforcing bodies.

D. Vasilev (✉) · E. Kravets
Volgograd Academy of the Russian Ministry of Internal
Affairs, Volgograd 400089, Russia
e-mail: 89889599848@mail.ru

E. Kravets
e-mail: 80kravez@gmail.com

Y. Naumov
Academy of Management of the Ministry of Internal
Affairs of Russia, Moscow 125993, Russian Federation
e-mail: Naumov6112@rambler.ru

E. Bulgakova
Kutafin Moscow State Law University, Moscow 125993, Russia
e-mail: oblaka7777777@gmail.com

V. Bulgakov
Moscow University of the Ministry of Internal Affairs of the Russian
Federation Named After V.Y. Kikot, Moscow 117437, Russia
e-mail: vg.bulgakov@mail.ru

© Springer Nature Switzerland AG 2019
A. G. Kravets (ed.), *Big Data-driven World: Legislation Issues and Control
Technologies*, Studies in Systems, Decision and Control 181,
https://doi.org/10.1007/978-3-030-01358-5_22

Intellectual monitoring of oil products creates conditions for preventing crimes in fuel and energy complex.

Keywords Data array · Information · Analysis of accumulated data
Theft in the fuel and energy complex · Crime analysis · Intellectual monitoring

1 Introduction

As noted by leading legal scholarship specializing in the study of crime in the oil and gas industry P. V. Biketov, in particular "… the dynamics of production and sales on foreign and local markets of the products of the oil and gas industry of the Russian Federation is currently one of the main factors determining the state of the country's economy. Further development of the economy to a large extent depends on the state of the industry infrastructure, the availability of equipment and mechanisms and the level of security. At the same time, the constantly growing demand for petroleum products and the reliability of investments make the sphere of oil refining, supply and petroleum products sales very attractive for obtaining both legal and criminal proceeds. It is economic attractiveness that determines the criminalization of the economic activity methods practically at any technological stage: from the extraction of hydrocarbon energy carriers and the production of finished products up to their sale" [1].

At the same time, the situation in the oil and gas industry is characterized by a steady growth of the criminal impact on the economy of enterprises that are part of this industry, which is primarily dangerous by erosion of their economic security. Since, as noted, the energy sector is one of the key elements of the country's budget, this creates a negative impact on the overall level of economic and, therefore, national security in general.

2 The Criminal Influence on the Energy Sector

The definition of the sphere of social relations as an object of scientific knowledge has an important criminological significance, as it summarizes its integrity, independence, specific architectonics of interelement links, integration into a more general system (national economy), the stability of contradictions, imbalances and crisis phenomena that arise here. In the criminological sense, this means the existence of a single mechanism of causing harm to individual components and the object as a whole, the relative homogeneity of the crimes committed in this sphere [2], the typical determinants that generate and contribute to the spread of the crime under consideration.

First, the sphere of economic crimes applies to all the types of economic activity related to the cost of energy (fuel and electricity), its creation, exchange, distribution and consumption in the production relations system. In other words, for this category of crime, the sphere of criminal influence is the national energy market.

Secondly, since the fuel industry is an organically integrated interindustry production complex, combined with production links of economic entities, a closed technological cycle, centralized management, vertical integration into the national economic complex, the criminal influence on this sector can be considered systemic, because on the national scale the way of production of energy carriers is violated, the means of production and industrial relations are harmed, which in general threatens the energy security of the state [3, 4]. The systemic nature of the criminal impact on the fuel and energy complex is reflected in the general mechanism of misappropriation of material goods, violation of economic interests of commodity producers, transporters, and consumers of energy resources.

Thirdly, the generic object of criminal influence in the sphere under consideration is the system of economic relations that cover technical-economic and organizational-economic relations that are formed in the process of production and consumption of energy carriers and the management of this process. The criminal encroachments in the fuel industry as a whole are directed against the mechanism of organization and functioning of the national economy in the sphere of energy production and consumption [5, 6]. The grading on the direct objects of causing harm provides for the division of the economic crimes aggregation in the fuel industry according to the component elements of this economic mechanism:

I.   encroachments on the property complex of enterprises and industrial products;
II. encroachments that violate the procedure for carrying out entrepreneurial activities in the energy market;
III. encroachments that violate the settlement terms for energy carriers;
IV. encroachments that violate the procedure for administrative and management activities implementation of public authorities and bodies in the implementation of the government policy in the fuel industry. The unifying basis for this type of crime is its affiliation with the sphere of economic relations, including property relations, economic and management relations, as well as the dominant lucrative motivation associated with unlawful enrichment and other personal benefits [7]. From the above data, it is possible to generalize the specific nature of the crimes of the category under consideration, since they are covered by a single economic process in the production of energy carriers and are committed in the course of carrying out industrial, economic, financial and official activities.

At the same time, in addition to encroachment on the main direct object, theft from oil pipelines, petroleum product pipelines and gas pipelines can harm simultaneously another object (other objects). In criminal law, such crimes are called two-object or multi-object, depending on how many additional direct objects the criminal act simultaneously causes harm to. It should be emphasized that in case of theft from oil pipelines, petroleum product pipelines and gas pipelines, in addition to the violation of the main direct object, the damage is usually caused additionally to public relations that ensure public safety [8, 9]. This distinctive feature of the criminal delicts under consideration was taken into account by the legislator when constructing the norms of the Special Part of the Criminal Code of Russia with the introduction of Article 215.3, which provides amenability for the destruction, damage or otherwise making inoperable the condition of pipelines, as well as technologically related facilities, communications, automation, alarm systems that caused or could lead to their malfunctioning and were committed from mercenary or hooligan motives. In accordance with this normative provision, theft of oil and petroleum products from oil pipelines and petroleum product pipelines associated with causing damage to pipeline transport facilities (criminal "tie-ins") entail responsibility for the totality of crimes provided for under item "b" of Part 3 of Art. 158 and the corresponding part of Art. 215.3 of the Criminal Code.

Since, as was shown above, theft of hydrocarbons belongs to the sphere of economic crime, the methods of combating and preventing this type of crimes must correspond to the general direction of combating economic crimes, including taking into account international experience [10–12].

The law enforcement practice of recent years shows that the oil and gas enterprises, like no others, are subject to unlawful attacks at the stage of hydrocarbon production and transportation. Considering that they are mainly transported by pipeline transport, the legislator directly calls the theft of hydrocarbons from the pipeline, committed through a criminal tie-in, the main type of criminal encroachment.

In general, the encroachment on the property complex of enterprises of the fuel and energy industry and their industrial products does not cover the economic (entrepreneurial) activity of these enterprises with property and financial assets as such. Analysis of the accumulated data allows us to state that theft of hydrocarbons belongs to the group of mercenary crimes connected with: (a) theft of hydrocarbon raw materials and products of its processing during extraction, processing, transportation, distribution and storage; (b) dismantling on scrap and hijacking industrial equipment, power units, automation equipment, communication, technological protection, equipment. Theft of hydrocarbon raw materials and products of its processing, on the one hand, is carried out massively and spontaneously with the participation of industry workers, and on the other hand, and, as a rule, on a professional basis, organized criminal groups. Among such crimes, thefts of oil and gas condensate are most common through damage and unauthorized connections (tie-ins) to main pipelines. This kind of criminal activity is put on an industrial basis and has acquired signs of organized oil and gas business, as the technology of abduction, pumping oil and gas raw materials, transportation, storage, processing in clandestine shops and selling petroleum products in the consumer market is established.

The issues of preventing individual crimes against property in the oil and gas industry were addressed in the monographs of the researchers [1, 4, 5].

However, complex studies devoted to the criminological characteristics and the development on its basis of the prevention of theft of hydrocarbons are not sufficient for the full development of the topic of this study.

Based on the annual reports of the security service of PJSC "Transneft" (full name "Public Joint Stock Company." If you look at the Charter, the name of this organization is Public Joint Stock Company "Transneft"), it can be concluded that formally, the fall in the number of crimes related to the theft of carbohydrates from pipeline transport is observed.

So, if in 2008—387 crimes of this orientation were registered, then in 2011—275, in 2012—313, in 2013—214, in 2014—180, in 2015—174, and in 2016 and 2017—185 cases. At the same time, the leading regions in which the level of theft of oil and petroleum products from petroleum product pipelines has remained at a high level in recent years are the Samara region, the Saratov region, the Irkutsk region, the Moscow region, the Republic of Dagestan.

So, according to the information presented at the meeting, for the year 2017, 35 facts of oil thefts were registered at the facilities of "Transneft-Privolga" Joint-Stock Company, including 33 unauthorized tie-ins (in 2016—25 tie-ins), as well as two thefts of oil from technological devices in the regions where the enterprise is present. Out of 33 facts of unauthorized tie-ins, in 19 cases they were manufactured with a tap.

In 2015, in the Samara region, 57 criminal cases were initiated on the facts of unauthorized tie-ins, in 2014—58. The largest number of unauthorized tie-ins was recorded in the territory of the Samara region—20 or 60% of the total number of tie-ins recorded in the regions where the enterprise is present.

At the same time, in order to assess the true degree of the public danger of the investigated acts, it is necessary to analyze the whole array of data on the damage caused by the crime. According to various expert estimates, on average more than five million tons of oil is stolen annually in Russia, the damage is more than $800 million. In particular, only up to 130,000 tons of oil is stolen from Transneft's pipelines every year. To this should be added the cost of repair and restoration and environmental restoration works. The total economic losses from oil thefts according to "VTB-Capital" are estimated in the range of 55–106 billion rubles, and 19–37 billion for the federal budget [13].

3   The Systems and Methods for Locating Leaks in Pipelines

It should be noted that in most cases, the presented data do not only demonstrate a real decrease in crime, but also an actual decrease in the number of crimes actually registered, since criminals today use the latest technologies so that the activities aimed at stealing hydrocarbons are not obvious for law enforcement agencies and enterprise security services. In particular, criminals use modern high-tech equipment for installation works, use horizontal drilling equipment, CCTV cameras in the area of illegal work, etc.

There are many systems and methods for locating leaks in pipelines that were designed to search for emergency leaks. But most systems cannot determine the occurrence of an unauthorized tie-in or sabotage, which is due to the limitations of the methods they are based on. Those systems that can identify both unauthorized tie-ins and sabotage require huge investments for their implementation. At present, the issue of monitoring the state of pipelines is relevant, with the cost as low as possible in real time.

Pipeline systems are one of the most cost-effective and safe ways of transporting gases, oil, petroleum products and other liquids. As a means of transportation over long distances, pipelines have a high degree of safety, reliability, and efficiency.

Most of the pipelines, regardless of the medium being transported, are developed based on a service life of about 25 years. As they age, they begin to fail, leaks in the structural weak points of the joints, corrosion points and areas with small structural damage to the material appeared. In addition, there are other causes leading to leakage, such as accidental damage to the pipeline, terrorist acts, sabotage, theft of the product from the pipeline, etc.

The main task of leak detection systems (LDS) is to help the pipeline owner to identify the leak and to determine its location. The LDS provides the generation of an alarm signal about the possible leakage and displays information that helps to decide whether there are leaks or not. The systems for detecting leaks from pipelines are of great importance for the operation of pipelines since they allow to reduce the pipeline downtime.

The term "leak detection system" and the LDS abbreviation is generally well established (used in a number of corporate regulatory documents of "Transneft Joint-stock company" OJSC). A number of manufacturers use different names for the designation of such systems:

1. Leak Detection and Activity Control System (LD and ACS)—"Omega" CJSC
2. Infrasound Pipeline Monitoring System (IPMS)—"Tori" SPC.

In English-speaking community practice, this kind of system is commonly called the Leak detection system (LDS).

The most common classification of LDS is given in Standard 1130, developed by the API (American Petroleum Institute).

According to this classification, LDS are subdivided into systems based on processes occurring in the pipeline and LDS on the basis of processes occurring outside the pipeline. Systems of the first kind use control equipment (pressure sensors, flowmeters, temperature sensors, etc.) to monitor the parameters of the transported medium in the pipeline. Systems of the second kind use control equipment (IR radiometers, thermal imagers, vapor detectors, acoustic microphones, fiber optic sensors, etc.) for monitoring parameters outside the pipeline.

A more particular classification is contained in RD-13.320.00-KTN-223-09 "Leak Detection Systems of a Combined Type on Main Oil Pipelines. General technical requirements for designing, manufacturing and commissioning", which is developed and applied by "Transneft Joint-stock company" OJSC. This classification covers only some of the systems considered in API 1130 as a system based on processes occurring in the pipeline. According to it, the LDS are divided into the following types:

1. Pressure wave leak detection system is a software and hardware complex for detecting a pressure wave that arises in the pipeline when a leak is formed in it.

The work of the complex is based on the analysis by the specialized software of data collected by specialized controllers (modules) of the LDS from additional (not used for process control) pressure sensors.

2. Parametric leak detection system is a software complex that operates in conjunction with the monitoring and supervisory control system (MSCS) based on the use of data on pipeline operation parameters entering the MSCS. The work of the complex is based on the analysis of the telemetry data available at the upper level of the automated process control system and the application of a mathematical model to make a decision on the presence of a leak. Systems of this kind in API 1155 are called "Software Based Leak Detection Systems";

3. Combined leak detection systems—LDS that combine a leakage detection system with a pressure wave and a parametric leak detection system.

In particular, the introduction of an administrative and legal regime for the registration and protection of fuel and energy resources contributes to the prevention of theft of hydrocarbons in the extraction, production, supply, transportation, storage and use of hydrocarbons.

Within the framework of the proposed regime, it is advisable to oblige natural monopolies and related markets (producers and suppliers) of fuel and energy resources to equip oil pipelines, gas distribution stations, gas pipelines and gas storages with automated accounting systems of transported products, which will allow in an operative mode to control the volume of hydrocarbon raw materials and products of their processing, with technically permissible errors to calculate the amount of damage, determine the places and areas of excessive losses, and also make it impossible to create unaccounted balances of energy sources [14].

At the same time, there is a need to introduce a consolidated and sectoral balance of fuel and energy resources at the disposal of energy companies, supply companies, and transport networks. An important measure preventing the theft of oil and petroleum products is the introduction of standardization and normalization of unit costs and losses in their production, transportation, and supply to consumers.

Unlike the regime accounting measures, indirectly preventing the theft of fuel and energy resources at various stages of the technological chain, security measures create direct obstacles to the execution of relevant criminal activities.

In particular, the following protective measures can contribute to improving the effectiveness of preventing theft of oil and gas condensate through damage and unauthorized connection to the main oil, gas, and petroleum product pipelines:

- coordination between the transporting organizations and the Ministry of Internal Affairs of the Russian Federation of the list of pipeline transport facilities, the protection of which should be carried out by departmental security services or subdivisions of the internal affairs bodies on a contractual basis and a joint initiative for legislative enactment on fixing such list with a regulatory act;
- approval by the leaders of the transport companies of the provisions (instructions) on the order of protection of oil and condensate pipelines, which, among other things, would provide for the duty and algorithms of interaction of officials with regional police units in cases where oil, gas condensate and oil and gas theft is

revealed [15, 16], as well as inspections of police records of persons who are appointed to departmental security units or to posts, official duties which are connected with material responsibility;
- carrying out measures of technical and physical protection of the objects of fuel and energy complex.

Technical protection measures provide for the equipment of transport mains with security devices, various kinds of sensors for acoustic, vibration and other engineering and technical activities that can immediately reveal the facts of unauthorized connections to pipelines and determine their localization [17]. On the most criminally vulnerable areas, unmanned aerial vehicles, helicopters, warning systems and even space monitoring of pipeline integrity can be used. The physical protection measures are complementary to the aforementioned, and mainly involve the development of patrolling routes by the relevant police units, as well as the positioning of patrols, arrangement of camouflaged observation posts, well-thought-out stationing of rapid reaction teams, in order to take prompt measures to detect and reduce illegal activities harm from it.

In addition to the abovementioned measures, the effectiveness of preventing and stopping the theft of fuel and energy resources directly depends on the timely detection of criminal intentions on the part of persons who conceive and are preparing for the commission of such crimes.

In connection with this, the operational and preventive work of criminal investigation units, police units are needed to collect, verify, record, accumulate, process and use operational information to identify and prosecute members of organized criminal groups that commit theft, storage, processing of hydrocarbons and marketing of products of their processing [18, 19]. In particular, it is advisable for criminal investigation officers to focus on collecting information about the way of life, the circle of people who: (a) have highly professional skills of gas welders; (b) have cars with tanks; (c) work at gas stations or are owners of such facilities; (d) are unemployed or have only been released from places of deprivation of liberty and do not have legal sources of income. Employees of the State Road Safety Inspectorate, whose right to inspect vehicles with tanks for fuels and lubricants transportation, should focus on the collection of data on the type, volume, data of cargo owners, transportation route (points of loading and delivery), and fundamental data for carriers. It is advisable to put the posts and patrols of the traffic police on the road sections in the areas of oil depots, oil and gas processing plants (especially small oil and gas processing enterprises). Personnel of Main Directorate for Economic Safety and Counteracting Corruption should focus on collecting data on individuals who are small suppliers of raw materials for oil and gas processing enterprises, including mini-factories. For maximum efficiency, it should be ensured that the collected information in the form of highly standardized reports is transmitted to the data bank of the information system of the internal affairs body [20], and the software of this system allows integrating the downloaded information and performing an automated search for necessary data.

Among the preventive measures, an effective place is occupied by the work aimed at reducing demand in the shadow market of oil and gas raw materials. In particular, it is a question of revealing and termination of the activity of underground oil refining mini-factories, as well as of the bases of storage of finished oil and gas products.

To narrow the sales markets for stolen oil and gas products, it is necessary to solve the problem of accrediting laboratories and equipping them with the appropriate equipment, which will allow us to carry out expert studies of quality of the fuel sold by gas stations in order to identify the facts of the sale of low-quality products to consumers and to bring to justice the guilty persons and illegal activity of such filling stations [21].

4 Conclusions

So, summing up the above, it should be noted that theft of oil and petroleum products refers to the crimes that threaten the entire national security system of Russia, due to the strategic importance of the oil and gas complex for the economic security of the state.

In the domestic scientific literature, relying on the requirements of paragraph "b" part 3 of Art. 158 of the Criminal Code of the Russian Federation, in our opinion, the form and method of committing the theft of oil and petroleum products are narrowed. Thus, theft can be committed openly, with the use of violence or weapons, which does not allow it to be regarded as a theft, also it can be carried out in other ways, apart from tapping into the oil and gas pipeline: by forgery of official documents, thefts from railroad or truck tanks, etc.

The unifying generic object sign of this type of crime is, undoubtedly, that theft of hydrocarbons is related to economic crimes, which makes it necessary to involve administrative methods in combating them and preventing them from spreading, including those built on the interaction of the organizations engaged in the transportation of oil and petroleum products, and law enforcement agencies.

Taking into account the high technological level and the real industrial scale of the type of crime being considered, the most effective measures to combat them lie in the development of modern automated leakage monitoring systems from pipelines and related software.

References

1. Biketov, P.V.: Criminological measures to counter theft of oil and petroleum products, committed at enterprises of the oil and gas industry: the author's abstract, p. 30. Thesis work. Candidate in Juridical Sciences, Moscow (2010)
2. Zhang, H.L., Liu, C.X., Zhao, M.Z.: Economics, fundamentals, technology, finance, speculation and geopolitics of crude oil prices: an econometric analysis and forecast based on data from 1990 to 2017. Pet. Sci. **15**, 432 (2018). https://doi.org/10.1007/s12182-018-0228-z
3. Li, H., Zhang, H.M., Xie, Y.T.: Analysis of factors influencing the Henry Hub natural gas price based on factor analysis. Pet. Sci. **14**, 822 (2017). https://doi.org/10.1007/s12182-017-0192-z

4. Piza, E.L., Gilchrist, A.M., Caplan, J.M., et al.: The financial implications of merging proactive CCTV monitoring and directed police patrol: a cost-benefit analysis. ExpCriminol **12**, 403 (2016). https://doi.org/10.1007/s11292-016-9267-x
5. Borzenkov, G.: Crimes Against Property. Criminal Law Course: Particular, vol. 3, p. 417. Moscow, IKD Zertsalo-M (2002)
6. Tseloni, A.: Exploring the international decline in crime rates. Eur. J. Criminol. **7**(5), 375–394 (2010)
7. Guerette, R., Clarke, R.: Product life cycles and crime: automated teller machines and robbery. Secur. J. **16**, 7 (2003). https://doi.org/10.1057/palgrave.sj.8340122
8. Levi, M., Morgan, J., Burrows, J.: Enhancing business crime reduction: UK directors responsibilities to review the impact of crime on business. Secur. J. **16**, 7 (2003). https://doi.org/10.1057/palgrave.sj.8340143
9. Kravets, E., et al.: Cognitive activity efficiency factors during investigative actions, performed using information and communication technologies. In: Communications in Computer and Information Science, Knowledge-Based Software Engineering: 11th Joint Conference, JCKBSE-2014, Volgograd, Russia, 17–20 Sept, t. 466, pp. 585–592 (2014)
10. Tilley, N., Hopkins, M.: Organized crime and local businesses. Criminol. Crim. Justice **8**(4), 443–459 (2008)
11. İşeri, E.: Addressing pipeline security regime of the prospective regional energy hub Turkey. Secur. J. **28**, 1 (2015). https://doi.org/10.1057/sj.2012.38
12. Ashby, M.P.J.: Is metal theft committed by organized crime groups, and why does it matter? Criminol. Crim. Justice **16**(2), 141–157 (2016)
13. RBC information agency. Official site URL: http://top.rbc.ru/economics/28/01/2013/842358.shtml. Accessed 28 June 2018
14. Sprott, J.B., Sutherland, J.: Unintended consequences of multiple bail conditions for youth. Can. Crim. Justice Assoc. **57**(1), 59–83 (2015)
15. Bradshaw, E.A.: "Obviously, we're all oil industry": the criminogenic structure of the offshore oil industry. Theor. Criminol. **19**(3), 376–395 (2015)
16. Koziratskii, Y.L., Popelo, V.D., Rusinov, P.S.: A method of prompt determination for points of unauthorized connection to major oil product pipelines. Chem. Petrol Eng. **35**, 150 (1999). https://doi.org/10.1007/BF02368196
17. Ngada, T., Bowers, K.: Spatial and temporal analysis of crude oil theft in the Niger delta. Secur. J. **31**, 501 (2018). https://doi.org/10.1057/s41284-017-0112-3
18. Edwards, A., Gill, P.: Crime as enterprise? The case of "transnational organisedcrime". Crime Law Soc. Change **37**, 203 (2002). https://doi.org/10.1023/A:1015025509582
19. Ma, W.J., Wei, Y.Z., Tao, S.Z.: A method for evaluating paleo hydrocarbon pools and predicting secondary reservoirs: a case study of the Sangonghe formation in the Mosuowan area, Junggar Basin. Pet. Sci. **15**, 252 (2018). https://doi.org/10.1007/s12182-018-0231-4
20. Buell, S.W.: The responsibility gap in corporate crime. Crim. Law Philos. https://doi.org/10.1007/s11572-017-9434-9 (2017)
21. de Vasconcellos Araujo, M., et al.: Hydrodynamic study oil leakage in pipeline via CFD. Adv. Mech. Eng. **6**, 170–178 (2014)

Author Index

© Springer Nature Switzerland AG 2019
A. G. Kravets (ed.), *Big Data-driven World: Legislation Issues and Control Technologies*, Studies in Systems, Decision and Control 181,
https://doi.org/10.1007/978-3-030-01358-5